Learn

Eureka Math®
Grade 5
Modules 1 & 2

Published by Great Minds®.

Copyright © 2018 Great Minds®.

Printed in the U.S.A.
This book may be purchased from the publisher at eureka-math.org.
1 2 3 4 5 6 7 8 9 10 BAB 25 24 23 22

ISBN 978-1-64054-071-2

G5-M1-M2-L-05.2018

Learn • Practice • Succeed

Eureka Math® student materials for *A Story of Units®* (K–5) are available in the *Learn, Practice, Succeed* trio. This series supports differentiation and remediation while keeping student materials organized and accessible. Educators will find that the *Learn, Practice,* and *Succeed* series also offers coherent—and therefore, more effective—resources for Response to Intervention (RTI), extra practice, and summer learning.

Learn

Eureka Math Learn serves as a student's in-class companion where they show their thinking, share what they know, and watch their knowledge build every day. *Learn* assembles the daily classwork—Application Problems, Exit Tickets, Problem Sets, templates—in an easily stored and navigated volume.

Practice

Each *Eureka Math* lesson begins with a series of energetic, joyous fluency activities, including those found in *Eureka Math Practice*. Students who are fluent in their math facts can master more material more deeply. With *Practice,* students build competence in newly acquired skills and reinforce previous learning in preparation for the next lesson.

Together, *Learn* and *Practice* provide all the print materials students will use for their core math instruction.

Succeed

Eureka Math Succeed enables students to work individually toward mastery. These additional problem sets align lesson by lesson with classroom instruction, making them ideal for use as homework or extra practice. Each problem set is accompanied by a Homework Helper, a set of worked examples that illustrate how to solve similar problems.

Teachers and tutors can use *Succeed* books from prior grade levels as curriculum-consistent tools for filling gaps in foundational knowledge. Students will thrive and progress more quickly as familiar models facilitate connections to their current grade-level content.

Students, families, and educators:

Thank you for being part of the *Eureka Math*® community, where we celebrate the joy, wonder, and thrill of mathematics.

In the *Eureka Math* classroom, new learning is activated through rich experiences and dialogue. The *Learn* book puts in each student's hands the prompts and problem sequences they need to express and consolidate their learning in class.

What is in the Learn book?

Application Problems: Problem solving in a real-world context is a daily part of *Eureka Math*. Students build confidence and perseverance as they apply their knowledge in new and varied situations. The curriculum encourages students to use the RDW process—Read the problem, Draw to make sense of the problem, and Write an equation and a solution. Teachers facilitate as students share their work and explain their solution strategies to one another.

Problem Sets: A carefully sequenced Problem Set provides an in-class opportunity for independent work, with multiple entry points for differentiation. Teachers can use the Preparation and Customization process to select "Must Do" problems for each student. Some students will complete more problems than others; what is important is that all students have a 10-minute period to immediately exercise what they've learned, with light support from their teacher.

Students bring the Problem Set with them to the culminating point of each lesson: the Student Debrief. Here, students reflect with their peers and their teacher, articulating and consolidating what they wondered, noticed, and learned that day.

Exit Tickets: Students show their teacher what they know through their work on the daily Exit Ticket. This check for understanding provides the teacher with valuable real-time evidence of the efficacy of that day's instruction, giving critical insight into where to focus next.

Templates: From time to time, the Application Problem, Problem Set, or other classroom activity requires that students have their own copy of a picture, reusable model, or data set. Each of these templates is provided with the first lesson that requires it.

Where can I learn more about Eureka Math resources?

The Great Minds® team is committed to supporting students, families, and educators with an ever-growing library of resources, available at eureka-math.org. The website also offers inspiring stories of success in the *Eureka Math* community. Share your insights and accomplishments with fellow users by becoming a *Eureka Math* Champion.

Best wishes for a year filled with aha moments!

Jill Diniz

Jill Diniz
Director of Mathematics
Great Minds

The Read–Draw–Write Process

The *Eureka Math* curriculum supports students as they problem-solve by using a simple, repeatable process introduced by the teacher. The Read–Draw–Write (RDW) process calls for students to

1. Read the problem.

2. Draw and label.

3. Write an equation.

4. Write a word sentence (statement).

Educators are encouraged to scaffold the process by interjecting questions such as

- What do you see?

- Can you draw something?

- What conclusions can you make from your drawing?

The more students participate in reasoning through problems with this systematic, open approach, the more they internalize the thought process and apply it instinctively for years to come.

Contents

Module 1: Place Value and Decimal Fractions

Module 2: Multi-Digit Whole Number and Decimal Fraction Operations

Topic G: Partial Quotients and Multi-Digit Decimal Division

Topic H: Measurement Word Problems with Multi-Digit Division

Grade 5
Module 1

Farmer Jim keeps 12 hens in every coop. If Farmer Jim has 20 coops, how many hens does he have in all? If every hen lays 9 eggs on Monday, how many eggs will Farmer Jim collect on Monday? Explain your reasoning using words, numbers, or pictures.

_____

_____

_____

_____

Read   **Draw**   **Write**

Lesson 1:   Reason concretely and pictorially using place value understanding to relate adjacent base ten units from millions to thousandths.

© 2018 Great Minds®. eureka-math.org

3

Name _Fernanda_

Date _____

1. Use the place value chart and arrows to show how the value of the each digit changes. The first one has been done for you.

   a.   3.452 × 10 = ____34.52____

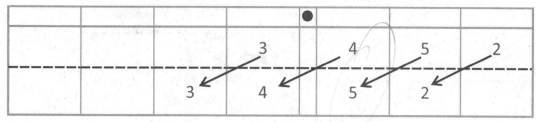

   b.   3.452 × 100 = _3 4 5 . 2_

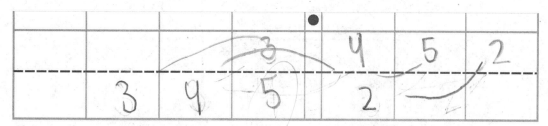

   c.   3.452 × 1,000 = _3452_

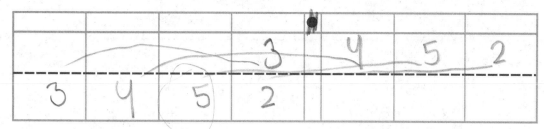

   d.   Explain how and why the value of the 5 changed in (a), (b), and (c).

   The value changed because I keep changing the place value

EUREKA
MATH

Lesson 1:   Reason concretely and pictorially using place value understanding to relate adjacent base ten units from millions to thousandths.

© 2018 Great Minds®. eureka-math.org

5

2. Use the place value chart and arrows to show how the value of each digit changes. The first one has been done for you.

   a. 345 ÷ 10 = ___34.5___

   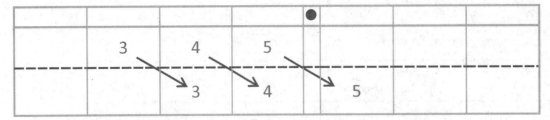

   b. 345 ÷ 100 = __3.45__

   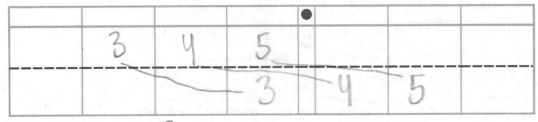

   c. 345 ÷ 1,000 = ___345___

   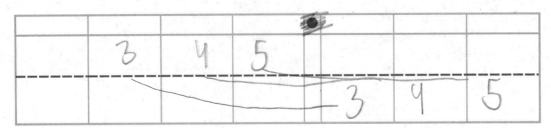

   d. Explain how and why the value of the 4 changed in the quotients in (a), (b), and (c).

Lesson 1:   Reason concretely and pictorially using place value understanding to relate adjacent base ten units from millions to thousandths.

EUREKA MATH®

3. A manufacturer made 7,234 boxes of coffee stirrers. Each box contains 1,000 stirrers. How many stirrers did they make? Explain your thinking, and include a statement of the solution.

4. A student used his place value chart to show a number. After the teacher instructed him to multiply his number by 10, the chart showed 3,200.4. Draw a picture of what the place value chart looked like at first.

Explain how you decided what to draw on your place value chart. Be sure to include your reasoning about how the value of each digit was affected by the multiplication. Use words, pictures, or numbers.

5. A microscope has a setting that magnifies an object so that it appears 100 times as large when viewed through the eyepiece. If a tiny insect is 0.095 cm long, how long will the insect appear in centimeters through the microscope? Explain how you know.

Lesson 1:   Reason concretely and pictorially using place value understanding to
               relate adjacent base ten units from millions to thousandths.

© 2018 Great Minds®. eureka-math.org

7

Name _____   Date _____

Use the place value chart and arrows to show how the value of each digit changes.

a.   6.671 × 100 = _____

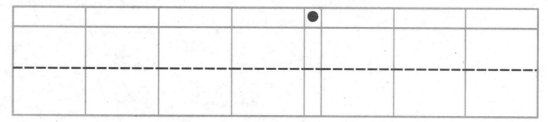

b.   684 ÷ 1,000 = _____

EUREKA
MATH®

Lesson 1:   Reason concretely and pictorially using place value understanding to
            relate adjacent base ten units from millions to thousandths.

© 2018 Great Minds®. eureka-math.org

9

A school district ordered 247 boxes of pencils. Each box contains 100 pencils. If the pencils are to be shared evenly among 10 classrooms, how many pencils will each class receive? Draw a place value chart to show your thinking.

Read   **Draw**   **Write**

Lesson 2:   Reason abstractly using place value understanding to relate adjacent
                 base ten units from millions to thousandths.   13

© 2018 Great Minds®. eureka-math.org

Name __Fernanda__   Date _____

1. Solve.

   a. 54,000 × 10 = __540,000__   e. 0.13 × 100 = __13.0__

   b. 54,000 ÷ 10 = __5,400__   f. 13 ÷ 1,000 = __.013__

   c. 8.7 × 10 = __87__   g. 3.12 × 1,000 = __3120__

   d. 8.7 ÷ 10 = __0.87__   h. 4,031.2 ÷ 100 = _____

2. Find the products.

   a. 19,340 × 10   = _____

   b. 19,340 × 100   = _____

   c. 19,340 × 1,000   = _____

   d. Explain how you decided on the number of zeros in the products for (a), (b), and (c).

EUREKA MATH®   Lesson 2:   Reason abstractly using place value understanding to relate adjacent
                                     base ten units from millions to thousandths.   15

© 2018 Great Minds®. eureka-math.org

3. Find the quotients.

    a.   152 ÷ 10 = _____

    b.   152 ÷ 100 = _____

    c.   152 ÷ 1,000 = _____

    d.   Explain how you decided where to place the decimal in the quotients for (a), (b), and (c).

4. Janice thinks that 20 hundredths is equivalent to 2 thousandths because 20 hundreds is equal to 2 thousands. Use words and a place value chart to correct Janice's error.

5. Canada has a population that is about $\frac{1}{10}$ as large as the United States. If Canada's population is about 32 million, about how many people live in the United States? Explain the number of zeros in your the answer.

Lesson 2:   Reason abstractly using place value understanding to relate adjacent base ten units from millions to thousandths.

EUREKA MATH

Name _____   Date _____

1. Solve.

    a. $32.1 \times 10 =$ _____

    b. $3632.1 \div 10 =$ _____

2. Solve.

    a. $455 \times 1,000 =$ _____

    b. $455 \div 1,000 =$ _____

$$\begin{array}{r} 50 \\ \times\ 11 \\ \hline 50 \end{array}$$

EUREKA MATH

Lesson 2:   Reason abstractly using place value understanding to relate adjacent
             base ten units from millions to thousandths.

© 2018 Great Minds®. eureka-math.org

17

Jack and Kevin are creating a mosaic for art class by using fragments of broken tiles. They want the mosaic to have 100 sections. If each section requires 31.5 tiles, how many tiles will they need to complete the mosaic? Explain your reasoning with a place value chart.

Read   **Draw**   **Write**

Lesson 3:   Use exponents to name place value units, and explain patterns in the placement of the decimal point.

19

© 2018 Great Minds®. eureka-math.org

$9 \times 10 \times 10 \times 10$

$39 \times 10 \times 10 \times 10 + 10$

Name _____ Date _____

1. Write the following in exponential form (e.g., $100 = 10^2$).

   a. $10,000 = \underline{10^4}$

   b. $1,000 = \underline{10^3}$

   c. $10 \times 10 = \underline{10^2}$

   d. $100 \times 100 = \underline{100^2}$

   e. $1,000,000 = \underline{10^{10}}$

   f. $1,000 \times 1,000 = \underline{1000^2}$

2. Write the following in standard form (e.g., $5 \times 10^2 = 500$).

   a. $9 \times 10^3 = \underline{9,000}$

   b. $39 \times 10^4 = \underline{\hphantom{xxxxxx}}$

   c. $7,200 \div 10^2 = \underline{\hphantom{xxxxxx}}$

   d. $7,200,000 \div 10^3 = \underline{\hphantom{xxxxxx}}$

   e. $4.025 \times 10^3 = \underline{\hphantom{xxxxxx}}$

   f. $40.25 \times 10^4 = \underline{\hphantom{xxxxxx}}$

   g. $72.5 \div 10^2 = \underline{\hphantom{xxxxxx}}$

   h. $7.2 \div 10^2 = \underline{\hphantom{xxxxxx}}$

3. Think about the answers to Problem 2(a–d). Explain the pattern used to find an answer when you multiply or divide a whole number by a power of 10.

4. Think about the answers to Problem 2(e–h). Explain the pattern used to place the decimal in the answer when you multiply or divide a decimal by a power of 10.

EUREKA MATH®

Lesson 3:   Use exponents to name place value units, and explain patterns in the placement of the decimal point.

© 2018 Great Minds®. eureka-math.org

21

5. Complete the patterns.

   a. 0.03   0.3   _____   30   _____   _____

   b. 6,500,000   65,000   _____   6.5   _____

   c. _____   9,430   _____   94.3   9.43   _____

   d. 999   9990   99,900   _____   _____   _____

   e. _____   7.5   750   75,000   _____   _____

   f. Explain how you found the unknown numbers in set (b). Be sure to include your reasoning about the number of zeros in your numbers and how you placed the decimal.

   g. Explain how you found the unknown numbers in set (d). Be sure to include your reasoning about the number of zeros in your numbers and how you placed the decimal.

6. Shaunnie and Marlon missed the lesson on exponents. Shaunnie incorrectly wrote $10^5 = 50$ on her paper, and Marlon incorrectly wrote $2.5 \times 10^2 = 2.500$ on his paper.

   a. What mistake has Shaunnie made? Explain using words, numbers, or pictures why her thinking is incorrect and what she needs to do to correct her answer.

   b. What mistake has Marlon made? Explain using words, numbers, or pictures why his thinking is incorrect and what he needs to do to correct his answer.

Lesson 3: Use exponents to name place value units, and explain patterns in the placement of the decimal point.

© 2018 Great Minds®. eureka-math.org

EUREKA MATH

Name _____   Date _____

1. Write the following in exponential form and as a multiplication sentence using only 10 as a factor (e.g., $100 = 10^2 = 10 \times 10$).

    a. 1,000   = _____ = _____

    b. 100×100 = _____ = _____

3. Write the following in standard form (e.g., $4 \times 10^2 = 400$).

    a. $3 \times 10^2 =$ _____

    b. $2.16 \times 10^4 =$ _____

    c. $800 \div 10^3 =$ _____

    d. $754.2 \div 10^2 =$ _____

EUREKA
MATH

Lesson 3:   Use exponents to name place value units, and explain patterns in the placement of the decimal point.

© 2018 Great Minds®. eureka-math.org

23

10	10 × _____	

_____

powers of 10 chart

Lesson 3:   Use exponents to name place value units, and explain patterns in the placement of the decimal point.

25

a. Use your meter strip to show and explain the length that relates to the hundredths and thousandths places. Record the results in the table.

thousands	hundreds	tens	ones	tenths	hundredths	thousandths
			1 meter	$\frac{1}{10}$ meter decimeter		

b. Explain the length that relates to the tens, hundreds, and thousands places. Record the results in the table.

Read   Draw   Write

Lesson 4:   Use exponents to denote powers of 10 with application to metric conversions.

© 2018 Great Minds®. eureka-math.org

27

Name _____   Date _____

1. Convert and write an equation with an exponent. Use your meter strip when it helps you.

   a. 3 meters to centimeters   3 m = 300 cm   _____ $3 \times 10^2 = 300$ _____

   b. 105 centimeters to meters   105 cm = _____ m   _____

   c. 1.68 meters to centimeters   _____ m = _____ cm   _____

   d. 80 centimeters to meters   _____ cm = _____ m   _____

   e. 9.2 meters to centimeters   _____ m = _____ cm   _____

   f. 4 centimeters to meters   _____ cm = _____ m   _____

   g. In the space below, list the letters of the problems where larger units are converted to smaller units.

2. Convert using an equation with an exponent. Use your meter strip when it helps you.

   a. 3 meters to millimeters   _____ m = _____ mm   _____

   b. 1.2 meters to millimeters   _____ m = _____ mm   _____

   c. 1,020 millimeters to meters   _____ mm = _____ m   _____

   d. 97 millimeters to meters   _____ mm = _____ m   _____

   e. 7.28 meters to millimeters   _____ m = _____ mm   _____

   f. 4 millimeters to meters   _____ mm = _____ m   _____

   g. In the space below, list the letters of the problems where smaller units are converted to larger units.

3. Read each aloud as you write the equivalent measures. Write an equation with an exponent you might use to convert.

    a. 3.512 m = _____ mm $3.512 \times 10^3 = 3,512$

    b. 8 cm = _____ m   _____

    c. 42 mm = _____ m   _____

    d. 0.05 m = _____ mm   _____

    e. 0.002 m = _____ cm   _____

4. The length of the bar for a high jump competition must always be 4.75 m. Express this measurement in millimeters. Explain your thinking. Include an equation with an exponent in your explanation.

5. A honey bee's length measures 1 cm. Express this measurement in meters. Explain your thinking. Include an equation with an exponent in your explanation.

6. Explain why converting from meters to centimeters uses a different exponent than converting from meters to millimeters.

Lesson 4:   Use exponents to denote powers of 10 with application to metric conversions.

© 2018 Great Minds®. eureka-math.org

EUREKA
MATH®

Name _____   Date _____

1. Convert using an equation with an exponent.

    a. 2 meters to centimeters   2 m = _____ cm   _____

    b. 40 millimeters to meters   40 mm = _____ m   _____

2. Read each aloud as you write the equivalent measures.

    a. A piece of fabric measures 3.9 meters. Express this length in centimeters.

    b. Ms. Ramos's thumb measures 4 centimeters. Express this length in meters.

Lesson 4:   Use exponents to denote powers of 10 with application to metric
               conversions.   31

© 2018 Great Minds®. eureka-math.org

Jordan measures a desk at 200 cm. James measures the same desk in millimeters, and Amy measures the same desk in meters. What is James's measurement in millimeters? What is Amy's measurement in meters? Show your thinking using a place value chart or an equation with exponents.

_____

_____

_____

Read   Draw   Write

Lesson 5:   Name decimal fractions in expanded, unit, and word forms by
                   applying place value reasoning.

© 2018 Great Minds®. eureka-math.org

33

Name _____   Date _____

1. Express as decimal numerals. The first one is done for you.

a.	Four thousandths	0.004
b.	Twenty-four thousandths	0.024
c.	One and three hundred twenty-four thousandths	1.324
d.	Six hundred eight thousandths	.608
e.	Six hundred and eight thousandths	600.008
f.	$\frac{46}{1000}$	0.046
g.	$3\frac{946}{1000}$	3.946
h.	$200\frac{904}{1000}$	200.904

2. Express each of the following values in words.

  a. 0.005   Five thousandths

  b. 11.037   eleven and thirty seven thousandths

  c. 403.608 four hundred three and six hundredth eight thousandth.

3. Write the number on a place value chart. Then, write it in expanded form using fractions or decimals to express the decimal place value units. The first one is done for you.

  a. 35.827

Tens	Ones		Tenths	Hundredths	Thousandths
3	5	●	8	2	7

$$35.827 = 3 \times 10 + 5 \times 1 + 8 \times \left(\frac{1}{10}\right) + 2 \times \left(\frac{1}{100}\right) + 7 \times \left(\frac{1}{1000}\right) \quad or$$
$$= 3 \times 10 + 5 \times 1 + 8 \times 0.1 + 2 \times 0.01 + 7 \times 0.001$$

    b.   0.249

    c.   57.281

4. Write a decimal for each of the following. Use a place value chart to help, if necessary.

    a. $7 \times 10 + 4 \times 1 + 6 \times \left(\frac{1}{10}\right) + 9 \times \left(\frac{1}{100}\right) + 2 \times \left(\frac{1}{1000}\right)$

    b. $5 \times 100 + 3 \times 10 + 8 \times 0.1 + 9 \times 0.001$

    c. $4 \times 1,000 + 2 \times 100 + 7 \times 1 + 3 \times \left(\frac{1}{100}\right) + 4 \times \left(\frac{1}{1000}\right)$

5. Mr. Pham wrote 2.619 on the board. Christy says it is two and six hundred nineteen thousandths. Amy says it is 2 ones 6 tenths 1 hundredth 9 thousandths. Who is right? Use words and numbers to explain your answer.

Lesson 5: Name decimal fractions in expanded, unit, and word forms by applying place value reasoning.

© 2018 Great Minds®. eureka-math.org

EUREKA
MATH®

Name _____ Date _____

1. Express nine thousandths as a decimal.

2. Express twenty-nine thousandths as a fraction.

3. Express 24.357 in words.

    a. Write the expanded form using fractions or decimals.

    b. Express in unit form.

Lesson 5:   Name decimal fractions in expanded, unit, and word forms by applying place value reasoning.

© 2018 Great Minds®. eureka-math.org

37

Thousandths	Hundredths	Tenths	Ones	Tens	Hundreds	Thousands

thousands through thousandths place value chart

Lesson 5:   Name decimal fractions in expanded, unit, and word forms by
              applying place value reasoning.

39

© 2018 Great Minds®. eureka-math.org

Ms. Meyer measured the edge of her dining table to the hundredths of a meter. The edge of the table measured 32.15 meters. Write her measurement in word form, unit form, and expanded form using fractions and decimals.

_____

_____

_____

_____

Read   Draw   Write

Lesson 6:   Compare decimal fractions to the thousandths using like units, and
express comparisons with >, <, =.

© 2018 Great Minds®. eureka-math.org

41

Name _____   Date _____

1. Show the numbers on the place value chart using digits. Use >, <, or = to compare. Explain your thinking in the space to the right.

34.223 (<) 34.232

	3	4	2	2	3
	3	4	2	3	2

0.8 (>) 0.706

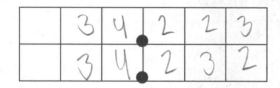

		0	8		
		0	7	0	6

2. Use >, <, or = to compare the following. Use a place value chart to help, if necessary.

a.	16.3	(<)	16.4
b.	0.83	(=)	$\frac{83}{100}$
c.	$\frac{205}{1000}$	(=)	0.205
d.	95.580	(>)	95.58
e.	9.1	(>)	9.099
f.	8.3	(>)	83 tenths
g.	5.8	(<)	Fifty-eight hundredths

h. Thirty-six and nine thousandths	◯	4 tens
i. 202 hundredths	◯	2 hundreds and 2 thousandths
j. One hundred fifty-eight thousandths	◯	158,000
k. 4.15	◯	415 tenths

3. Arrange the numbers in increasing order.

   a. 3.049   3.059   3.05   3.04

   _____

   b. 182.205   182.05   182.105   182.025

   _____

4. Arrange the numbers in decreasing order.

   a. 7.608   7.68   7.6   7.068

   _____

   b. 439.216   439.126   439.612   439.261

   _____

Lesson 6:   Compare decimal fractions to the thousandths using like units, and express comparisons with >, <, =.

EUREKA MATH®

5. Lance measured 0.485 liter of water.   Angel measured 0.5 liter of water. Lance said, "My beaker has more water than yours because my number has three decimal places and yours only has one." Is Lance correct? Use words and numbers to explain your answer.

6. Dr. Hong prescribed 0.019 liter more medicine than Dr. Tannenbaum. Dr. Evans prescribed 0.02 less than Dr. Hong. Who prescribed the most medicine? Who prescribed the least?

Lesson 6:   Compare decimal fractions to the thousandths using like units, and
express comparisons with >, <, =.

© 2018 Great Minds®. eureka-math.org

45

Name _____   Date _____

1. Show the numbers on the place value chart using digits. Use >, <, or = to compare. Explain your thinking in the space to the right.

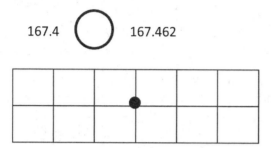

167.4 ◯ 167.462

2. Use >, <, and = to compare the numbers.

    32.725 ◯ 32.735

3. Arrange the numbers in decreasing order.

    76.342   76.332   76.232   76.343

    _____

EUREKA
MATH®

Lesson 6:   Compare decimal fractions to the thousandths using like units, and
express comparisons with >, <, =.

© 2018 Great Minds®. eureka-math.org

47

Craig, Randy, Charlie, and Sam ran in a 5K race on Saturday. They were the top 4 finishers. Here are their race times:

Craig: 25.9 minutes   Randy: 32.2 minutes

Charlie: 32.28 minutes   Sam: 25.85 minutes

Who won first place? Who won second place? Third? Fourth?

Read   **Draw**   **Write**

EUREKA MATH

Lesson 7:   Round a given decimal to any place using place value understanding
              and the vertical number line.

49

© 2018 Great Minds®. eureka-math.org

Name _____   Date _____

Fill in the table, and then round to the given place. Label the number lines to show your work. Circle the rounded number.

1. 3.1

    a. Hundredths   b. Tenths   c. Tens

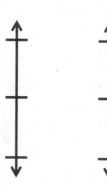

Tens	Ones	Tenths	Hundredths	Thousandths
		•		

2. 115.376

    a. Hundredths   b. Ones   c. Tens

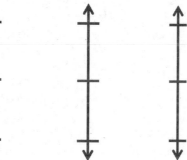

Tens	Ones	Tenths	Hundredths	Thousandths
		•		

EUREKA MATH®

Lesson 7:   Round a given decimal to any place using place value understanding and the vertical number line.

51

© 2018 Great Minds®. eureka-math.org

3. 0.994

Tens	Ones	Tenths	Hundredths	Thousandths
	●			

a. Hundredths   b. Tenths   c. Ones   d. Tens

4. For open international competition, the throwing circle in the men's shot put must have a diameter of 2.135 meters. Round this number to the nearest hundredth. Use a number line to show your work.

5. Jen's pedometer said she walked 2.549 miles. She rounded her distance to 3 miles. Her brother rounded her distance to 2.5 miles. When they argued about it, their mom said they were both right. Explain how that could be true. Use number lines and words to explain your reasoning.

Lesson 7:   Round a given decimal to any place using place value understanding and the vertical number line.

EUREKA
MATH

Name _____   Date _____

Use the table to round the number to the given places. Label the number lines, and circle the rounded value.

8.546

Tens	Ones	●	Tenths	Hundredths	Thousandths
	8	●	5	4	6
		●	85	4	6
		●		854	6
		●			8546

a. Hundredths

b. Tens

EUREKA
MATH®

Lesson 7: Round a given decimal to any place using place value understanding
and the vertical number line.

© 2018 Great Minds®. eureka-math.org

53

Hundreds	Tens	Ones	•	Tenths	Hundredths	Thousandths

hundreds to thousandths place value chart

Organic whole wheat flour sells in bags weighing 2.915 kilograms.

   a. How much flour is this when rounded to the nearest tenth? Use a place value chart and number line to explain your thinking.

   b. How much flour is this when rounded to the nearest one?

Read   **Draw**   **Write**

Lesson 8:   Round a given decimal to any place using place value understanding and the vertical number line.

© 2018 Great Minds®. eureka-math.org

57

Extension: What is the difference of the two answers?

Read   **Draw**   **Write**

Lesson 8:   Round a given decimal to any place using place value understanding and the vertical number line.

Name _____   Date _____

1. Write the decomposition that helps you, and then round to the given place value. Draw number lines to explain your thinking. Circle the rounded value on each number line.

    a. Round 32.697 to the nearest tenth, hundredth, and one.

    b. Round 141.999 to the nearest tenth, hundredth, ten, and hundred.

2. A root beer factory produces 132,554 cases in 100 days. About how many cases does the factory produce in 1 day? Round your answer to the nearest tenth of a case. Show your thinking on the number line.

Lesson 8:   Round a given decimal to any place using place value understanding and the vertical number line.

© 2018 Great Minds®. eureka-math.org

59

3. A decimal number has two digits to the right of its decimal point. If we round it to the nearest tenth, the result is 13.7.

    a. What is the maximum possible value of this number? Use words and the number line to explain your reasoning. Include the midpoint on your number line.

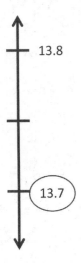

    b. What is the minimum possible value of this decimal? Use words and the number line to explain your reasoning. Include the midpoint on your number line.

Lesson 8: Round a given decimal to any place using place value understanding and the vertical number line.

© 2018 Great Minds®. eureka-math.org

EUREKA
MATH

Name _____   Date _____

Round the quantity to the given place value. Draw number lines to explain your thinking. Circle the rounded value on the number line.

a. 13.989 to the nearest tenth

b. 382.993 to nearest hundredth

Lesson 8:   Round a given decimal to any place using place value understanding
and the vertical number line.

© 2018 Great Minds®. eureka-math.org

61

Ten baseballs weigh 1,417.4 grams. About how much does 1 baseball weigh? Round your answer to the nearest tenth of a gram. Round your answer to the nearest gram. Which answer would you give if someone asked, "About how much does a baseball weigh?" Explain your choice.

_____

_____

_____

_____

Read   **Draw**   **Write**

Lesson 9:   Add decimals using place value strategies, and relate those strategies
to a written method.

63

© 2018 Great Minds®. eureka-math.org

Name _Fernanda-Arreaza_ Date _9/9/22_

1. Solve, and then write the sum in standard form. Use a place value chart if necessary.

   a. 1 tenth + 2 tenths = _____ tenths = _____

   b. 14 tenths + 9 tenths = _____ tenths = _____ one(s) _____ tenth(s) = _____

   c. 1 hundredth + 2 hundredths = _____ hundredths = _____

   d. 27 hundredths + 5 hundredths = _32_ hundredths = _3_ tenths _2_ hundredths = _0.32_

   e. 1 thousandth + 2 thousandths = _____ thousandths = _____

   f. 35 thousandths + 8 thousandths = ____ thousandths = ____ hundredths ____ thousandths = _____

   g. 6 tenths + 3 thousandths = _0.603_ thousandths = _____

   h. 7 ones 2 tenths + 4 tenths = _____ tenths = _____

   i. 2 thousandths + 9 ones 5 thousandths = _9.007_ thousandths = _____

2. Solve using the standard algorithm.

a. 0.3 + 0.82 = _____	b. 1.03 + 0.08 = _1.11_ $\begin{array}{r} 1 \\ +\ 1.03 \\ 0.008 \\ \hline 1.11 \end{array}$
c. 7.3 + 2.8 = _____	d. 57.03 + 2.08 = _59.11_ $\begin{array}{r} 57.03 \\ +\ 02.08 \\ \hline 59.11 \end{array}$

EUREKA MATH

Lesson 9: Add decimals using place value strategies, and relate those strategies to a written method.

© 2018 Great Minds®. eureka-math.org

65

e. 62.573 + 4.328 = _____	f. 85.703 + 12.197 = _____

3. Van Cortlandt Park's walking trail is 1.02 km longer than Marine Park's.   Central Park's walking trail is 0.242 km longer than Van Cortlandt's.

   a.   Fill in the missing information in the chart below.

$$
\begin{array}{r}
2.300 \\
+\,0.242 \\
\hline
2.542
\end{array}
$$

$$
\begin{array}{r}
+\,1.02 \\
1.28 \\
\hline
2.30
\end{array}
$$

New York City Walking Trails	
Central Park	2.542 km
Marine Park	→ 1.28 km
✱ Van Cortlandt Park	2.30 km

   b.   If a tourist walked all 3 trails in a day, how many kilometers would he or she have walked?

$$
\begin{array}{r}
2.542 \\
+\,1.280 \\
2.300 \\
\hline
6.122
\end{array}
$$

The tourist walked 6.122 Km.

4. Meyer has 0.64 GB of space remaining on his iPod. He wants to download a pedometer app (0.24 GB), a photo app (0.403 GB), and a math app (0.3 GB). Which combinations of apps can he download? Explain your thinking.

Lesson 9:   Add decimals using place value strategies, and relate those strategies to a written method.

EUREKA MATH®

Name __Fernanda – Arreaza__   Date __9|1|22__

1. Solve.

   a. 4 hundredths + 8 hundredths = __12__ hundredths = __1__ tenth(s) __2__ hundredths

   b. 64 hundredths + 8 hundredths = __72__ hundredths = __7__ tenths __2__ hundredths

2. Solve using the standard algorithm.

a. 2.40 + 1.8 = __4.20__	b. 36.25 + 8.67 = __44.92__
$$\begin{array}{r} 1 \\ 2.40 \\ +1.80 \\ \hline 4.20 \end{array}$$	$$\begin{array}{r} 1 \\ 36.25 \\ +8.67 \\ \hline 44.92 \end{array}$$

$$\begin{array}{r} \overset{1}{6}7 \\ +6 \\ \hline 73 \end{array}$$

$$\begin{array}{r} 26.27 \\ +8.68 \\ \hline 34.85 \end{array}$$

At the 2012 London Olympics, Michael Phelps won the gold medal in the men's 100-meter butterfly. He swam the first 50 meters in 26.96 seconds. The second 50 meters took him 25.39 seconds. What was his total time?

_____

_____

_____

_____

Read   **Draw**   **Write**

Lesson 10:   Subtract decimals using place value strategies, and relate those
strategies to a written method.

© 2018 Great Minds®. eureka-math.org

69

Name _____   Date _____

1. Subtract, writing the difference in standard form. You may use a place value chart to solve.

    a. 5 tenths – 2 tenths = _____ tenths = _____

    b. 5 ones 9 thousandths – 2 ones = _____ ones _____ thousandths = _____

    c. 7 hundreds 8 hundredths – 4 hundredths = _____ hundreds _____ hundredths = _____

    d. 37 thousandths – 16 thousandths = _____ thousandths = _____

2. Solve using the standard algorithm.

a. 1.4 – 0.7 = _____	b. 91.49 – 0.7 = _____	c. 191.49 – 10.72 = _____
d. 7.148 – 0.07 = _____	e. 60.91 – 2.856 = _____	f. 361.31 – 2.841 = _____

EUREKA MATH

Lesson 10:   Subtract decimals using place value strategies, and relate those strategies to a written method.

© 2018 Great Minds®. eureka-math.org

71

3. Solve.

a. 10 tens – 1 ten 1 tenth	b. 3 – 22 tenths	c. 37 tenths – 1 one 2 tenths
d. 8 ones 9 hundredths – 3.4	e. 5.622 – 3 hundredths	f. 2 ones 4 tenths – 0.59

4. Mrs. Fan wrote *5 tenths minus 3 hundredths* on the board. Michael said the answer is 2 tenths because 5 minus 3 is 2. Is he correct? Explain.

5. A pen costs $2.09. It costs $0.45 less than a marker. Ken paid for one pen and one marker with a five-dollar bill. Use a tape diagram with calculations to determine his change.

Lesson 10: Subtract decimals using place value strategies, and relate those strategies to a written method.

© 2018 Great Minds®. eureka-math.org

EUREKA MATH

Name _____ Date _____

1. Subtract.

    1.7 – 0.8 = _____ tenths – _____ tenths = ___ _____ tenths = _____

2. Subtract vertically, showing all work.

    a. 84.637 – 28.56 = _____

    b. 7 – 0.35 = _____

Lesson 10:   Subtract decimals using place value strategies, and relate those
strategies to a written method.

© 2018 Great Minds®. eureka-math.org

73

After school, Marcus ran 3.2 km, and Cindy ran 1.95 km. Who ran farther? How much farther?

Read   **Draw**   **Write**

Lesson 11:   Multiply a decimal fraction by single-digit whole numbers, relate to a
written method through application of the area model and place value
understanding, and explain the reasoning used.

© 2018 Great Minds®. eureka-math.org

75

Name _____   Date _____

1. Solve by drawing disks on a place value chart. Write an equation, and express the product in standard form.

    a. 3 copies of 2 tenths   b. 5 groups of 2 hundredths

    c. 3 times 6 tenths   d. 6 times 4 hundredths

    e. 5 times as much as 7 tenths   f. 4 thousandths times 3

2. Draw a model similar to the one pictured below for Parts (b), (c), and (d). Find the sum of the partial products to evaluate each expression.

    a. 7×3.12 3 ones   +   1 tenth   +   2 hundredths

7	7×3 ones	7×1 tenth	7×2 hundredths

                _____   +   _____   +   0.14   = _____

    b. 6×4.25

EUREKA
MATH

Lesson 11:   Multiply a decimal fraction by single-digit whole numbers, relate to a
             written method through application of the area model and place value
             understanding, and explain the reasoning used.
© 2018 Great Minds®. eureka-math.org

77

c. 3 copies of 4.65

d. 4 times as much as 20.075

3. Miles incorrectly gave the product of 7 × 2.6 as 14.42. Use a place value chart or an area model to help Miles understand his mistake.

4. Mrs. Zamir wants to buy 8 protractors and some erasers for her classroom. She has $30. If protractors cost $2.65 each, how much will Mrs. Zamir have left to buy erasers?

Lesson 11: Multiply a decimal fraction by single-digit whole numbers, relate to a written method through application of the area model and place value understanding, and explain the reasoning used.

© 2018 Great Minds®. eureka-math.org

EUREKA
MATH

Name _____   Date _____

1. Solve by drawing disks on a place value chart. Write an equation, and express the product in standard form.

   4 copies of 3 tenths

2. Complete the area model, and then find the product.

   3×9.63

   _____   _____   _____

   _____ | $3 \times$ ____ ones | $3 \times$ ____ tenths | $3 \times$ ____ hundredths |

EUREKA MATH

Lesson 11:   Multiply a decimal fraction by single-digit whole numbers, relate to a written method through application of the area model and place value understanding, and explain the reasoning used.

© 2018 Great Minds®. eureka-math.org

79

Patty buys 7 juice boxes a month for lunch. If one juice box costs $2.79, how much money does Patty spend on juice each month? Use an area model to solve.

Extension: How much will Patty spend on juice in 10 months? In 12 months?

_____

_____

_____

_____

Read   **Draw**   **Write**

EUREKA MATH

Lesson 12:   Multiply a decimal fraction by single-digit whole numbers, including using estimation to confirm the placement of the decimal point.

© 2018 Great Minds®. eureka-math.org

81

Name _____   Date_____

1. Choose the reasonable product for each expression. Explain your reasoning in the spaces below using words, pictures, or numbers.

   a. 2.5 × 4   0.1   1   10   100

   _____

   b. 3.14 × 7   2198   219.8   21.98   2.198

   _____

   c. 8 × 6.022   4.8176   48.176   481.76   4817.6

   _____

   d. 9 × 5.48   493.2   49.32   4.932   0.4932

Lesson 12:   Multiply a decimal fraction by single-digit whole numbers, including
             using estimation to confirm the placement of the decimal point.

© 2018 Great Minds®. eureka-math.org   83

2. Pedro is building a spice rack with 4 shelves that are each 0.55 meter long. At the hardware store, Pedro finds that he can only buy the shelving in whole meter lengths. Exactly how many meters of shelving does Pedro need? Since he can only buy whole-number lengths, how many meters of shelving should he buy? Justify your thinking.

3. Marcel rides his bicycle to school and back on Tuesdays and Thursdays. He lives 3.62 kilometers away from school. Marcel's gym teacher wants to know about how many kilometers he bikes in a week. Marcel's math teacher wants to know exactly how many kilometers he bikes in a week. What should Marcel tell each teacher? Show your work.

4. The poetry club had its first bake sale, and they made $79.35. The club members are planning to have 4 more bake sales. Leslie said, "If we make the same amount at each bake sale, we'll earn $3,967.50." Peggy said, "No way, Leslie! We'll earn $396.75 after five bake sales." Use estimation to help Peggy explain why Leslie's reasoning is inaccurate. Show your reasoning using words, numbers, or pictures.

EUREKA MATH

Name _____   Date _____

1. Use estimation to choose the correct value for each expression.

    a. 5.1×2 0.102   1.02   10.2   102

    b. 4×8.93 3.572   35.72   357.2   3572

2. Estimate the answer for 7.13×6. Explain your reasoning using words, pictures, or numbers.

EUREKA
MATH

Lesson 12:   Multiply a decimal fraction by single-digit whole numbers, including
using estimation to confirm the placement of the decimal point.

85

© 2018 Great Minds®. eureka-math.org

Louis buys 4 chocolates. Each chocolate costs $2.35. Louis multiplies 4 × 235 and gets 940. Place the decimal to show the cost of the chocolates, and explain your reasoning using words, numbers, and pictures.

_____

_____

_____

_____

Read   **Draw**   **Write**

Name _____   Date _____

1. Complete the sentences with the correct number of units, and then complete the equation.

   a. 4 groups of _____ tenths is 1.6. $1.6 \div 4 =$ _____

   b. 8 groups of _____ hundredths is 0.32. $0.32 \div 8 =$ _____

   c. 7 groups of _____ thousandths is 0.084. $0.084 \div 7 =$ _____

   d. 5 groups of _____ tenths is 2.0. $2.0 \div 5 =$ _____

2. Complete the number sentence. Express the quotient in units and then in standard form.

   a. $4.2 \div 7 =$ _____ tenths $\div 7 =$ _____ tenths $=$ _____

   b. $2.64 \div 2 =$ _____ ones $\div 2 +$ _____ hundredths $\div 2$

      $=$ _____ ones $+$ _____ hundredths

      $=$ _____

   c. $12.64 \div 2 =$ _____ ones $\div 2 +$ _____ hundredths $\div 2$

      $=$ _____ ones $+$ _____ hundredths

      $=$ _____

   d. $4.26 \div 6 =$ _____ tenths $\div 6 +$ _____ hundredths $\div 6$

      $=$ _____

      $=$ _____

EUREKA
MATH®

Lesson 13:   Divide decimals by single-digit whole numbers involving easily
             identifiable multiples using place value understanding and relate
             to a written method.

© 2018 Great Minds®. eureka-math.org

89

e. $4.236 \div 6 =$ _____

= _____

= _____

3.   Find the quotients. Then, use words, numbers, or pictures to describe any relationships you notice between each pair of problems and quotients.

a. $32 \div 8 =$ _____ $3.2 \div 8 =$ _____

b. $81 \div 9 =$ _____ $0.081 \div 9 =$ _____

4.   Are the quotients below reasonable? Explain your answers.

a. $5.6 \div 7 = 8$

b. $56 \div 7 = 0.8$

c. $0.56 \div 7 = 0.08$

Lesson 13:   Divide decimals by single-digit whole numbers involving easily identifiable multiples using place value understanding and relate to a written method.

© 2018 Great Minds®. eureka-math.org

EUREKA
MATH®

5. 12.48 milliliters of medicine were separated into doses of 4 mL each. How many doses were made?

6. The price of milk in 2013 was around $3.28 a gallon. This was eight times as much as you would have probably paid for a gallon of milk in the 1950s. What was the cost for a gallon of milk during the 1950s? Use a tape diagram, and show your calculations.

Name _____   Date _____

1. Complete the sentences with the correct number of units, and then complete the equation.

    a. 2 groups of _____ tenths is 1.8. $1.8 \div 2 =$ _____

    b. 4 groups of _____ hundredths is 0.32. $0.32 \div 4 =$ _____

    c. 7 groups of _____ thousandths is 0.021. $0.021 \div 7 =$ _____

2. Complete the number sentence. Express the quotient in unit form and then in standard form.

    a. $4.5 \div 5 =$ _____ tenths $\div 5 =$ _____ tenths = _____

    b. $6.12 \div 6 =$ _____ ones $\div 6\ +$ _____ hundredths $\div 6$

        $=$ _____ ones + _____ hundredths

        $=$ _____

A bag of potato chips contains 0.96 grams of sodium. If the bag is split into 8 equal servings, how many grams of sodium will each serving contain?

Extension: What other ways can the bag be divided into equal servings so that the amount of sodium in each serving has two digits to the right of the decimal and the digits are greater than zero in the tenths and hundredths place?

_____

_____

_____

_____

Read   **Draw**   **Write**

Name _____ Date _____

1. Draw place value disks on the place value chart to solve. Show each step using the standard algorithm.

a. 4.236 ÷ 3 = _1412_

Ones	Tenths	Hundredths	Thousandths
9	2	3	6
1	4	1	2

$$
\begin{array}{r}
1412 \\
3\,\overline{)4\,.\,2\,3\,6} \\
-3 \\
\overline{12} \\
-12 \\
\overline{03} \\
-3 \\
\overline{06}
\end{array}
$$

b. 1.324 ÷ 2 = _____

Ones	Tenths	Hundredths	Thousandths

$$
2\,\overline{)1\,.\,3\,2\,4}
$$

EUREKA
MATH®

Lesson 14: Divide decimals with a remainder using place value understanding
and relate to a written method.

97

© 2018 Great Minds®. eureka-math.org

2. Solve using the standard algorithm.

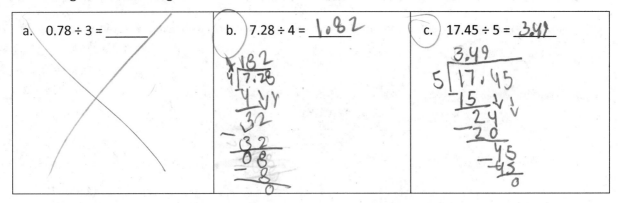

a. $0.78 \div 3 =$ _____

b. $7.28 \div 4 =$ 1.82

c. $17.45 \div 5 =$ 3.49

3. Grayson wrote $1.47 \div 7 = 2.1$ in her math journal.
   Use words, numbers, or pictures to explain why Grayson's thinking is incorrect.

4. Mrs. Nguyen used 1.48 meters of netting to make 4 identical mini hockey goals. How much netting did she use per goal?

5. Esperanza usually buys avocados for $0.94 apiece. During a sale, she gets 5 avocados for $4.10. How much money did she save per avocado? Use a tape diagram, and show your calculations.

EUREKA MATH

Name _____   Date _____

1. Draw place value disks on the place value chart to solve. Show each step using the standard algorithm.

   5.372 ÷ 2 = _____

Ones	Tenths	Hundredths	Thousandths

   $$2\overline{)5.372}$$

2. Solve using the standard algorithm.

   0.576 ÷ 4 = _____

Lesson 14:   Divide decimals with a remainder using place value understanding and relate to a written method.

© 2018 Great Minds®. eureka-math.org

99

Jose bought a bag of 6 oranges for $2.82. He also bought 5 pineapples. He gave the cashier $20 and received $1.43 change. How much did each pineapple cost?

_____

_____

_____

_____

Read   **Draw**   **Write**

Lesson 15:   Divide decimals using place value understanding, including remainders in the smallest unit.

101

Name _____   Date _____

1. Draw place value disks on the place value chart to solve. Show each step in the standard algorithm.

   a.   0.5 ÷ 2 = _____

Ones	•	Tenths	Hundredths	Thousandths

$$2\overline{)0.5}$$

   b.   5.7 ÷ 4 = _____

Ones	•	Tenths	Hundredths	Thousandths

$$4\overline{)5.7}$$

EUREKA MATH

Lesson 15:   Divide decimals using place value understanding, including remainders in the smallest unit.

103

© 2018 Great Minds®. eureka-math.org

2. Solve using the standard algorithm.

a. $0.9 \div 2 =$	b. $9.1 \div 5 =$	c. $9 \div 6 =$
d. $0.98 \div 4 =$	e. $9.3 \div 6 =$	f. $91 \div 4 =$

3. Six bakers shared 7.5 kilograms of flour equally. How much flour did they each receive?

4. Mrs. Henderson makes punch by mixing 10.9 liters of apple juice, 0.6 liters of orange juice, and 8 liters of ginger ale. She pours the mixture equally into 6 large punch bowls. How much punch is in each bowl? Express your answer in liters.

Lesson 15: Divide decimals using place value understanding, including remainders in the smallest unit.

© 2018 Great Minds®. eureka-math.org

EUREKA MATH

Name _____   Date _____

1. Draw place value disks on the place value chart to solve. Show each step in the standard algorithm.

   0.9 ÷ 4 = _____

Ones	•	Tenths	Hundredths	Thousandths

$$4\overline{)0.\ 9}$$

2. Solve using the standard algorithm.

   9.8 ÷ 5 =

Lesson 15:   Divide decimals using place value understanding, including remainders in the smallest unit.

105

© 2018 Great Minds®. eureka-math.org

Jesse and three friends buy snacks for a hike. They buy trail mix for $5.42, apples for $2.55, and granola bars for $3.39. If the four friends split the cost of the snacks equally, how much should each friend pay?

_____

_____

_____

_____

Read   **Draw**   **Write**

Name _____ Date _____

Solve.

1. Mr. Frye distributed $126 equally among his 4 children for their weekly allowance.

   a. How much money did each child receive?

   b. John, the oldest child, paid his siblings to do his chores. If John pays his allowance equally to his brother and two sisters, how much money will each of his siblings have received in all?

2. Ava is 23 cm taller than Olivia, and Olivia is half the height of Lucas. If Lucas is 1.78 m tall, how tall are Ava and Olivia? Express their heights in centimeters.

3. Mr. Hower can buy a computer with a down payment of $510 and 8 monthly payments of $35.75. If he pays cash for the computer, the cost is $699.99. How much money will he save if he pays cash for the computer instead of paying for it in monthly payments?

4. Brandon mixed 6.83 lb of cashews with 3.57 lb of pistachios. After filling up 6 bags that were the same size with the mixture, he had 0.35 lb of nuts left. What was the weight of each bag? Use a tape diagram, and show your calculations.

Lesson 16: Solve word problems using decimal operations.

EUREKA MATH

5. The bakery bought 4 bags of flour containing 3.5 kg each. 0.475 kg of flour is needed to make a batch of muffins, and 0.65 kg is needed to make a loaf of bread.

a. If 4 batches of muffins and 5 loaves of bread are baked, how much flour will be left? Give your answer in kilograms.

b. The remaining flour is stored in bins that hold 3 kg each. How many bins will be needed to store the flour? Explain your answer.

Name _____   Date _____

Write a word problem with two questions that matches the tape diagram below, and then solve.

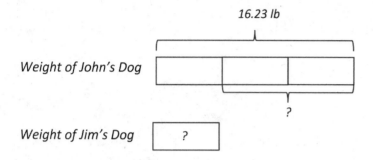

16.23 lb

Weight of John's Dog

?

Weight of Jim's Dog   ?

Grade 5
Module 2

The top surface of a desk has a length of 5.6 feet. The length is 4 times its width. What is the width of the desk?

_____

_____

_____

_____

Read   **Draw**   **Write**

Lesson 1:   Multiply multi-digit whole numbers and multiples of 10 using place value patterns and the distributive and associative properties.

117

Name _____   Date _____

1. Fill in the blanks using your knowledge of place value units and basic facts.

a. 23 × 20	b. 230 × 20
Think: 23 ones × 2 tens = __46__ tens	Think: 23 tens × 2 tens = __460__
23 × 20 = __460__	230 × 20 = __4600__
c. 41 × 4	d. 410 × 400
41 ones × 4 ones = 164 __tens__	41 tens × 4 hundreds = 164 __thousands__
41 × 4 = __164__	410 × 400 = __16,000__
e. 3,310 × 300	f. 500 × 600
_____ tens × _____ hundreds = 993 _____	_____ hundreds × _____ hundreds = 30 _____
3,310 × 300 = _____	500 × 600 = _____

2. Determine if these equations are true or false. Defend your answer using your knowledge of place value and the commutative, associative, and/or distributive properties.

   a. 6 tens = 2 tens × 3 tens

   b. 44 × 20 × 10 = 440 × 2

   c. 86 ones × 90 hundreds = 86 ones × 900 tens

   d. 64 × 8 × 100 = 640 × 8 × 10

Lesson 1:   Multiply multi-digit whole numbers and multiples of 10 using place
            value patterns and the distributive and associative properties.

119

© 2018 Great Minds®. eureka-math.org

e. $57 \times 2 \times 10 \times 10 \times 10 = 570 \times 2 \times 10$

3.   Find the products. Show your thinking. The first row gives some ideas for showing your thinking.

a. 7×9 7×90 70×90 70×900
     $= 63$ $= 63 \times 10$ $= (7 \times 10) \times (9 \times 10)$ $= (7 \times 9) \times (10 \times 100)$
                         $= 630$ $= (7 \times 9) \times 100$ $= 63,000$
                                            $= 6,300$

b. 45×3 45×30 450×30 450×300

c. 40×5 40×50 40×500 $400 \times 5,000$

d. 718×2 $7,180 \times 20$ $7,180 \times 200$ $71,800 \times 2,000$

Lesson 1:   Multiply multi-digit whole numbers and multiples of 10 using place
value patterns and the distributive and associative properties.

EUREKA
MATH

4. Ripley told his mom that multiplying whole numbers by multiples of 10 was easy because you just count zeros in the factors and put them in the product. He used these two examples to explain his strategy.

7,000 × 600 = 4,200,000   800 × 700 = 560,000
(3 zeros) (2 zeros) (5 zeros)   (2 zeros) (2 zeros) (4 zeros)

Ripley's mom said his strategy will not always work. Why not? Give an example.

5. The Canadian side of Niagara Falls has a flow rate of 600,000 gallons per second. How many gallons of water flow over the falls in 1 minute?

6. Tickets to a baseball game are $20 for an adult and $15 for a student. A school buys tickets for 45 adults and 600 students. How much money will the school spend for the tickets?

Lesson 1: Multiply multi-digit whole numbers and multiples of 10 using place value patterns and the distributive and associative properties.

© 2018 Great Minds®. eureka-math.org

121

Name _____ Date _____

1. Find the products.

    a. 1,900 × 20

    b. 6,000 × 50

    c. 250 × 300

$$
\begin{array}{r}
1,900 \\
\times\ \ \ 20 \\
\hline
000 \\
+1800\ \ \\
2000\ \ \ \\
\hline
3\,800
\end{array}
$$

$$
\begin{array}{r}
6000 \\
\times\ \ \ 50 \\
\hline
+\ 000 \\
30000 \\
\hline
300,000
\end{array}
$$

2. Explain how knowing 50 × 4 = 200 helps you find 500 × 400.

EUREKA
MATH

Lesson 1:   Multiply multi-digit whole numbers and multiples of 10 using place value patterns and the distributive and associative properties.

123

© 2018 Great Minds®. eureka-math.org

$\frac{1}{1,000}$	Thousandths					
$\frac{1}{100}$	Hundredths					
$\frac{1}{10}$	Tenths					
•	•	•	•	•	•	•
1	Ones					
10	Tens					
100	Hundreds					
1,000	Thousands					
10,000	Ten Thousands					
100,000	Hundred Thousands					
1,000,000	Millions					

millions to thousandths place value chart

Lesson 1:   Multiply multi-digit whole numbers and multiples of 10 using place
value patterns and the distributive and associative properties.

125

Jonas practices guitar 1 hour a day for 2 years. Bradley practices the guitar 2 hours a day more than Jonas. How many more minutes does Bradley practice the guitar than Jonas over the course of 2 years?

_____

_____

_____

_____

Read   **Draw**   **Write**

Lesson 2:   Estimate multi-digit products by rounding factors to a basic fact and
using place value patterns.

© 2018 Great Minds®. eureka-math.org

127

Name **Fernanda**   Date **9/21/2022**

1. Round the factors to estimate the products.

   a. $597 \times 52 \approx$ __600__ $\times$ __50__ $=$ __30,000__

   A reasonable estimate for 597×52 is __30,000__.

   b. $1,103 \times 59 \approx$ __1,000__ $\times$ __60__ $=$ __60,000__

   A reasonable estimate for $1,103 \times 59$ is __60,000__.

   c. $5,840 \times 25 \approx$ __6,000__ $\times$ __30__ $=$ __180,000__

   A reasonable estimate for $5,840 \times 25$ is __180,000__.

2. Complete the table using your understanding of place value and knowledge of rounding to estimate the product.

Expressions	Rounded Factors	Estimate
a. $2,809 \times 42$	$3,000 \times 40$	✓ 120,000
b. $28,090 \times 420$		
c. $8,932 \times 59$		
d. 89 tens $\times$ 63 tens		
e. 398 hundreds $\times$ 52 tens		

EUREKA MATH

Lesson 2: Estimate multi-digit products by rounding factors to a basic fact and using place value patterns.

© 2018 Great Minds®. eureka-math.org

129

3. For which of the following expressions would 200,000 be a reasonable estimate? Explain how you know.

   2,146 × 12   21,467 × 121   2,146 × 121   21,477 × 1,217

4. Fill in the missing factors to find the given estimated product.

   a. 571 × 43 ≈ _____ × _____ = 24,000

   b. 726 × 674 ≈ _____ × _____ = 490,000

   c. 8,379 × 541 ≈ _____ × _____ = 4,000,000

5. There are 19,763 tickets available for a New York Knicks home game.   If there are 41 home games in a season, about how many tickets are available for all the Knicks' home games?

6. Michael saves $423 dollars a month for college.

   a. About how much money will he have saved after 4 years?

   b. Will your estimate be lower or higher than the actual amount Michael will save? How do you know?

Lesson 2:   Estimate multi-digit products by rounding factors to a basic fact and using place value patterns.

© 2018 Great Minds®. eureka-math.org

EUREKA
MATH®

Name _Fernanda_____ Date _9/21/2022____

Round the factors and estimate the products.

a. $656 \times 106 \approx$ 700 × 100 = 70,000

b. $3,108 \times 7,942 \approx$ 3,000 × 8,000 = 24,000,000

c. $425 \times 9,311 \approx$ 400 × 9,000 = 360,0000

d. $8,633 \times 57,008 \approx$ 9,000 × 60,000 = 540,000,000

Lesson 2: Estimate multi-digit products by rounding factors to a basic fact and using place value patterns.

131

© 2018 Great Minds®. eureka-math.org

Robin is 11 years old. Her mother, Gwen, is 2 years more than 3 times Robin's age.
How old is Gwen?

_____

_____

_____

_____

Read   **Draw**   **Write**

Lesson 3: Write and interpret numerical expressions, and compare expressions
using a visual model.

© 2018 Great Minds®. eureka-math.org

133

Name <u>Fernanada</u>   Date _____

1. Draw a model. Then, write the numerical expressions.

a. The sum of 8 and 7, doubled	b. 4 times the sum of 14 and 26
$(8+7) \times 2 = 30$ 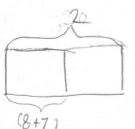 20 (8+7)	$4 \times (14) \times 26 =$ 4 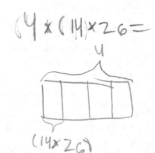 (14×26)
c. 3 times the difference between 37.5 and 24.5	d. The sum of 3 sixteens and 2 nines
$3 \times (37.5 - 24.5)$ 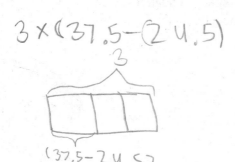 3 (37.5−24.5)	
e. The difference between 4 twenty-fives and 3 twenty-fives	f. Triple the sum of 33 and 27

2. Write the numerical expressions in words. Then, solve.

Expression	Words	The Value of the Expression
a. $12 \times (5 + 25)$	12 times (5 plus 25)	
b. $(62 - 12) \times 11$	(62 minus 12) times 11	
c. $(45 + 55) \times 23$		
d. $(30 \times 2) + (8 \times 2)$		

3. Compare the two expressions using >, <, or =. In the space beneath each pair of expressions, explain how you can compare without calculating. Draw a model if it helps you.

a. $24 \times (20 + 5)$	◯	$(20 + 5) \times 12$
b. 18×27 27×18 56 $+ 160$ 170 200 486	<	20 twenty-sevens minus 1 twenty-seven $\times 27$ 20 $(20 \times 27) - (1 \times 27)$ 140 400 540
c. 19×9 19 $\times 9$ 18 $+ 90$ 108	◯	3 nineteens, tripled

Lesson 3:   Write and interpret numerical expressions, and compare expressions using a visual model.

© 2018 Great Minds®. eureka-math.org

EUREKA MATH®

4. Mr. Huynh wrote *the sum of 7 fifteens and 38 fifteens* on the board.

   Draw a model, and write the correct expression.

5. Two students wrote the following numerical expressions.

       Angeline: $(7 + 15) \times (38 + 15)$

       MeiLing: $15 \times (7 + 38)$

   Are the students' expressions equivalent to your answer in Problem 4? Explain your answer.

6. A box contains 24 oranges. Mr. Lee ordered 8 boxes for his store and 12 boxes for his restaurant.

   a. Write an expression to show how to find the total number of oranges ordered.

   b. Next week, Mr. Lee will double the number of boxes he orders. Write a new expression to represent the number of oranges in next week's order.

   c. Evaluate your expression from Part (b) to find the total number of oranges ordered in both weeks.

Name _____ Date _____

1. Draw a model. Then, write the numerical expressions.

a. The difference between 8 forty-sevens and 7 forty-sevens	b. 6 times the sum of 12 and 8

2. Compare the two expressions using >, <, or =.

62 × (70 + 8)	◯	(70 + 8) × 26

EUREKA MATH

Lesson 3: Write and interpret numerical expressions, and compare expressions using a visual model.

© 2018 Great Minds®. eureka-math.org

139

Jaxon earned $39 raking leaves. His brother, Dayawn, earned 7 times as much waiting on tables.

Write a numerical expression to show Dayawn's earnings. How much money did Dayawn earn?

Read   **Draw**   **Write**

Lesson 4:   Convert numerical expressions into unit form as a mental strategy
for multi-digit multiplication.

141

© 2018 Great Minds®. eureka-math.org

Name _Fernanda_____   Date _____

1. Circle each expression that is not equivalent to the expression in **bold**.

    a. **16 × 29**

    29 sixteens   16 × (30 − 1)   (15 − 1) × 29   (10 × 29) − (6 × 29)

    b. **38 × 45**

    (38 + 40) × (38 + 5)   (38 × 40) + (38 × 5)   45 × (40 + 2)   45 thirty-eights

    c. **74 × 59**

    74 × (50 + 9)   74 × (60 − 1)   (74 × 5) + (74 × 9)   59 seventy-fours

2. Solve using mental math. Draw a tape diagram and fill in the blanks to show your thinking. The first one is partially done for you.

    a. 19 × 25 = __475__ twenty-fives

25	25	25	...	25	25
1	2	3	...	19	20

    Think: 20 twenty-fives − 1 twenty-five.

    = (__20__ × 25) − (__1__ × 25)

    = __500__ − __25__

    = __475__

    b. 24 × 11 = __11__ twenty-fours

    20 × 11   + 4

    Think: __20__ twenty fours + __4__ twenty four

    = (__20__ × 24) + (__4__ × 24)

    = __480__ + __96__

    = __576__

c. 79 × 14 = ____98____ fourteens

d. 21 × 75 = ____226____ seventy-fives

$$20$$

$$150$$
$$+\ \ 75$$
$$\overline{225}$$

$$75$$
$$\times\ \ 2$$
$$+\ \ 10$$
$$\overline{140}$$
$$\overline{150}$$

Think: ___80___ fourteens – 1 fourteen

= (__80__ × 14) – (___1___ × 14)

= ___112___ – ___14___

= ___98___

Think: ___20___ seventy-fives + ___1___ seventy-five

= (___20___ × 75) + (___1___ × 75)

= ___150___ + ___75___

= ___226___

3. Define the unit in word form and complete the sequence of problems as was done in the lesson.

a. 19 × 15 = 19 _____

Think: 20 _____ – 1 _____

= (20 × _____) – (1 × _____)

= _____ – _____

= _____

b. 14 × 15 = 14 _____

Think: 10 _____ + 4 _____

= (10 × _____) + (4 × _____)

= _____ + _____

= _____

EUREKA
MATH®

c.   25 × 12 = 12 _____

d.   18 × 17 = 18 _____

Think: 10 _____ + 2 _____

= (10 × _____) + (2 × _____)

= _____ + _____

= _____

Think: 20 _____ − 2 _____

= (20 × _____) − (2 × _____)

= _____ − _____

= _____

4.   How can 14 × 50 help you find 14 × 49?

5.   Solve mentally.

a.   101 × 15 = _____

b.   18 × 99 = _____

6.   Saleem says 45 × 32 is the same as (45 × 3) + (45 × 2). Explain Saleem's error using words, numbers, and/or pictures.

7.   Juan delivers 174 newspapers every day. Edward delivers 126 more newspapers each day than Juan.

a.   Write an expression to show how many newspapers Edward will deliver in 29 days.

b.   Use mental math to solve. Show your thinking.

Lesson 4:   Convert numerical expressions into unit form as a mental strategy
               for multi-digit multiplication.

145

© 2018 Great Minds®. eureka-math.org

Name __Fernanda_____   Date _____

Solve using mental math. Draw a tape diagram and fill in the blanks to show your thinking.

a. 49 × 11 = ___534___ elevens

b. 25 × 13 = ___325___ twenty-fives

$$
\begin{array}{r}
\times\ \frac{11}{5} \\
5 \\
+\ 5\ 0 \\
\hline
5\ 5
\end{array}
$$

+3

Think: 50 elevens − 1 eleven

= (__50__ × 11) − (__1__ × 11)

= __550__ − __11__

= __539__

Think: __10__ twenty-fives + __3__ twenty-fives

= (__10__ × 25) + (__3__ × 25)

= __250__ + __75__

= __325__

$$
\begin{array}{r}
25 \\
\times\ 1\ 3 \\
\hline
1\ 7\ 5 \\
2\ 5\ 0 \\
\hline
3\ 2\ 5
\end{array}
$$

LHT

EUREKA
MATH®

Lesson 4:   Convert numerical expressions into unit form as a mental strategy
             for multi-digit multiplication.

© 2018 Great Minds®. eureka-math.org

147

Aneisha is setting up a play space for her new puppy. She will be building a rectangular fence around part of her yard that measures 29 feet by 12 feet. How many square feet of play space will her new puppy have? If you have time, solve in more than one way.

Read   Draw   Write

Lesson 5:   Connect visual models and the distributive property to partial products
of the standard algorithm without renaming.

149

© 2018 Great Minds®. eureka-math.org

Name __Fernanda__   Date _____

1. Draw an area model, and then solve using the standard algorithm. Use arrows to match the partial products from the area model to the partial products of the algorithm.

a. $34 \times 21 = $ __714__

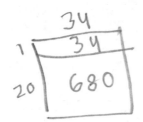

$$\begin{array}{r} 1 \\ 6\,8\,0 \\ \times 3\,4 \\ \hline 7\,1\,4 \end{array}$$

$$\begin{array}{r} 3\,4 \\ \times\ 2\,1 \\ \hline 4 \\ 3\,0 \\ +\ 8\,0 \\ \underline{-6\,0\,0} \\ 7\,1\,4 \end{array}$$

b. $434 \times 21 = $ __9114__

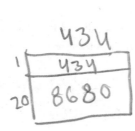

$$\begin{array}{r} 1\ \ 1 \\ 8{,}6\,8\,0 \\ +\ \ \ 4\,3\,4 \\ \hline 9{,}1\,1\,4 \end{array}$$

$$\begin{array}{r} 4\,3\,4 \\ \times\ \ 2\,1 \\ \hline 1434 \\ 81680 \\ \hline 9{,}1\,1\,4 \end{array}$$

2. Solve using the standard algorithm.

a. $431 \times 12 = $ __5172__

$$\begin{array}{r} 4\,3\,1 \\ \times\ 1\,2 \\ \hline 8\,6\,2 \\ 4\,3\,1\,0 \\ \hline 5{,}1\,7\,2 \end{array}$$

b. $123 \times 23 = $ _____

c. $312 \times 32 = $ __9,984__

$$\begin{array}{r} 3\,1\,2 \\ \times\ \ 3\,2 \\ \hline 6\,2\,4 \\ +\ 9{,}3\,6\,0 \\ \hline 9{,}9\,8\,4 \end{array}$$

EUREKA MATH   **Lesson 5:**   Connect visual models and the distributive property to partial products   **151**
of the standard algorithm without renaming.

© 2018 Great Minds®. eureka-math.org

3. Betty saves $161 a month. She saves $141 less each month than Jack. How much will Jack save in 2 years?

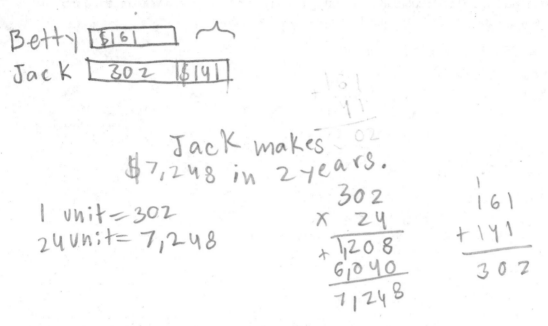

Betty [$161]

Jack [302 $141]

Jack makes
$7,248 in 2 years.

1 unit = 302
24 unit = 7,248

$$\begin{array}{r} 302 \\ \times\ 24 \\ \hline 1,208 \\ 6,040 \\ \hline 7,248 \end{array}$$

$$\begin{array}{r} 1 \\ 161 \\ +141 \\ \hline 302 \end{array}$$

4. Farmer Brown feeds 12.1 kilograms of alfalfa to each of his 2 horses daily. How many kilograms of alfalfa will all his horses have eaten after 21 days? Draw an area model to solve.

EUREKA
MATH®

Name _____   Date _____

Draw an area model, and then solve using the standard algorithm.

a. $21 \times 23 =$ ___483___

21

	21
3	63
20	420

$$\begin{array}{r} 420 \\ +\ 63 \\ \hline 483 \end{array}$$

$$\begin{array}{r} 2\ 1 \\ \times\ 2\ 3 \\ \hline +6\ 3 \\ 4\ 2\ 0 \\ \hline 4\ 8\ 3 \end{array}$$

b. $143 \times 12 =$ _____

10	430	

$$\begin{array}{r} 1\ 4\ 3 \\ \times\ \ \ 1\ 2 \\ \hline 2\ 8\ 6 \\ +\ 1\ 4\ 3\ 0 \\ \hline 1\ 7\ 1\ 6 \end{array}$$

EUREKA MATH®

Lesson 5:   Connect visual models and the distributive property to partial products of the standard algorithm without renaming.

© 2018 Great Minds®. eureka-math.org

153

Scientists are creating a material that may replace damaged cartilage in human joints. This *hydrogel* can stretch to 21 times its original length. If a strip of hydrogel measures 3.2 cm, what would its length be when stretched to capacity?

_____

_____

_____

_____

Read   **Draw**   **Write**

Lesson 6:   Connect area models and the distributive property to partial products
of the standard algorithm with renaming.

155

Name __Fernanda-Arreaza__   Date _____

1. Draw an area model. Then, solve using the standard algorithm. Use arrows to match the partial products from your area model to the partial products in the algorithm.

a. 48 × 35

```
        40      8
   ┌────────┬──────┐
 5 │  200   │  40  │
   ├────────┼──────┤
30 │ 1,200  │ 240  │
   └────────┴──────┘
```

```
  1,200
    200
+   240
     40
  ──────
  1,680
```

```
      4 8
  ×   3 5
  ─────────
       40
      240
  +   200
    1,200
  ─────────
    1,680
```

b. 648 × 35

```
        600     40     8
   ┌────────┬───────┬──────┐
 5 │ 3,000  │  200  │  40  │
   ├────────┼───────┼──────┤
30 │18,000  │ 1,200 │ 240  │
   └────────┴───────┴──────┘
```

```
 18,000
  3,000
 11,200
    240
    200
+    40
 ───────
 22,680
```

```
        4
      6 4 8
  ×     3 5
  ─────────
       200
    +3,000
    11,240
    18,000
  +    240
  ─────────
    22,680
```

EUREKA MATH

Lesson 6: Connect area models and the distributive property to partial products of the standard algorithm with renaming.

157

© 2018 Great Minds®. eureka-math.org

2. Solve using the standard algorithm.

a. 758 × 92

$$
\begin{array}{r}
758 \\
\times\ 92 \\
\hline
16 \\
100 \\
1,400 \\
720 \\
4,500 \\
63,000 \\
\hline
69,736
\end{array}
$$

b. 958 × 94

$$
\begin{array}{r}
958 \\
\times\ 94 \\
\hline
32 \\
2,200 \\
3,600 \\
720 \\
14,500 \\
81,000 \\
\hline
90,052
\end{array}
$$

c. 476 × 65

d. 547 × 64

3. Carpet costs $16 a square foot. A rectangular floor is 16 feet long by 14 feet wide. How much would it cost to carpet the floor?

Lesson 6:   Connect area models and the distributive property to partial products
              of the standard algorithm with renaming.

© 2018 Great Minds®. eureka-math.org

EUREKA
MATH®

4. General admission to The American Museum of Natural History is $19.

   a. If a group of 125 students visits the museum, how much will the group's tickets cost?

   b. If the group also purchases IMAX movie tickets for an additional $4 per student, what is the new total cost of all the tickets? Write an expression that shows how you calculated the new price.

Lesson 6: Connect area models and the distributive property to partial products
of the standard algorithm with renaming.

© 2018 Great Minds®. eureka-math.org

159

Name __Fernanda__   Date _____

Draw an area model. Then, solve using the standard algorithm. Use arrows to match the partial products from your area model to the partial products in the algorithm.

a.) 78 × 42

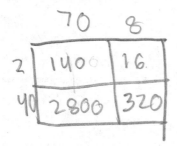

```
     1
    2,800
      140
  +   320
       16
    3,276
```

```
       78
     × 42
       16
      140
   +  320
     2800
    3,2 7 6
```

b.) 783 × 42

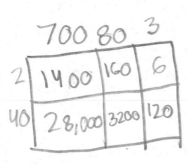

```
    1
   28,000
    3,200
    1,400
  +   160
      120
        6
   32,886
```

```
       783
     × 42
         6
       160
     1,400
       120
   +  3,200
     28,000
     321,886
```

The length of a school bus is 12.6 meters. If 9 school buses park end-to-end with 2 meters between each one, what's the total length from the front of the first bus to the end of the last bus?

_____

_____

_____

_____

Read   **Draw**   **Write**

Lesson 7:   Connect area models and the distributive property to partial products of the standard algorithm with renaming.

163

© 2018 Great Minds®. eureka-math.org

Name ___Fernanda~Arreaza___   Date _____

1. Draw an area model. Then, solve using the standard algorithm. Use arrows to match the partial products from the area model to the partial products in the algorithm.

   a. 481 × 352

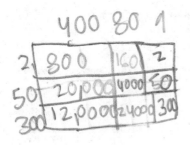

$$
\begin{array}{r}
1\ 20,000 \\
20,000 \\
24,000 \\
4,000 \\
+\quad 800 \\
300 \\
160 \\
50 \\
2 \\
\hline
169,312
\end{array}
$$

$$
\begin{array}{r}
481 \\
\times\ 352 \\
\hline
2 \\
160 \\
800 \\
50 \\
1\ 400 \\
20,000 \\
+\quad 300 \\
24,000 \\
1\ 20,000 \\
\hline
169,312
\end{array}
$$

   b. 481 × 302

$$
\begin{array}{r}
481 \\
\times\ 302 \\
\end{array}
$$

   c. Why are there three partial products in 1(a) and only two partial products in 1(b)?

2. Solve by drawing the area model and using the standard algorithm.

   a. 8,401 × 305

$$\begin{array}{r} 8,401 \\ \times\ \ \ \ 305 \\ \hline \end{array}$$

   b. 7,481 × 350

$$\begin{array}{r} 7,481 \\ \times\ \ \ \ 350 \\ \hline \end{array}$$

3. Solve using the standard algorithm.

   a. 346 × 27

   b. 1,346 × 297

EUREKA MATH

c. 346 × 207

d. 1,346 × 207

4. A school district purchased 615 new laptops for their mobile labs. Each computer cost $409. What is the total cost for all of the laptops?

5. A publisher prints 1,512 copies of a book in each print run. If they print 305 runs, how many books will be printed?

6. As of the 2010 census, there were 3,669 people living in Marlboro, New York. Brooklyn, New York, has 681 times as many people. How many more people live in Brooklyn than in Marlboro?

Lesson 7:   Connect area models and the distributive property to partial products of the standard algorithm with renaming.

167

© 2018 Great Minds®. eureka-math.org

Name _____   Date _____

Draw an area model. Then, solve using the standard algorithm.

a.   642 × 257

$$
\begin{array}{r}
6\,4\,2 \\
\times\ 2\,5\,7 \\
\hline
1\,4 \\
2\,8\,0 \\
4{,}2\,0\,0 \\
1\,0\,0 \\
2{,}0\,0\,0 \\
3\,0{,}0\,0\,0 \\
4\,0\,0 \\
2\,0{,}0\,0\,0 \\
1\,8\,0{,}0\,0\,0 \\
\hline
\end{array}
$$

236,994

b.   642 × 207

$$
\begin{array}{r}
6\,4\,2 \\
\times\ 2\,0\,7 \\
\hline
\end{array}
$$

Lesson 7:   Connect area models and the distributive property to partial products
of the standard algorithm with renaming.

169

EUREKA
MATH

© 2018 Great Minds®. eureka-math.org

Erin and Frannie entered a rug design contest. The rules stated that the rug's dimensions must be 32 inches × 45 inches and that they must be rectangular. They drew the following for their entries. Show at least three other designs they could have entered in the contest. Calculate the area of each section, and the total area of the rugs.

Erin

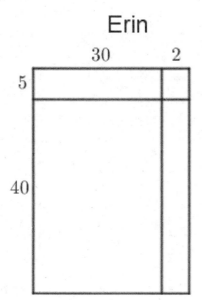

Frannie

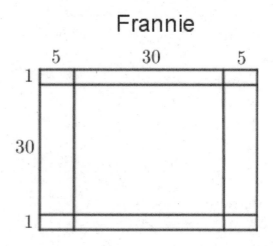

_____

_____

_____

_____

Read   Draw   Write

Lesson 8:   Fluently multiply multi-digit whole numbers using the standard algorithm and using estimation to check for reasonableness of the product.

© 2018 Great Minds®. eureka-math.org

171

Name __Fernanda-Arreaza__   Date __9/29/22__

1. Estimate the product first. Solve by using the standard algorithm. Use your estimate to check the reasonableness of the product.

a. 213 × 328	b. 662 × 372	c. 739 × 442
≈ 200 × 300 = 60,000 2 1 3 × 3 2 8		
d. 807 × 491	e. 3,502 × 656	f. 4.390 × 741
g. 530 × 2,075	h. 4,004 × 603	i. 987 × 3,105

g.
$$\begin{array}{r} 2075 \\ \times\ \ 530 \\ \hline +62,500 \\ 160,000 \\ 115,750 \\ \hline 1,273,250 \end{array}$$

$$\begin{array}{r} 2000 \\ \times\ 500 \\ \hline 10,4000,000 \end{array}$$

= 1,273,250

EUREKA
MATH

Lesson 8:   Fluently multiply multi-digit whole numbers using the standard algorithm and using estimation to check for reasonableness of the product.

© 2018 Great Minds®. eureka-math.org

173

2. Each container holds 1 L 275 mL of water. How much water is in 609 identical containers? Find the difference between your estimated product and precise product.

3. A club had some money to purchase new chairs. After buying 355 chairs at $199 each, there was $1,068 remaining. How much money did the club have at first?

Lesson 8:   Fluently multiply multi-digit whole numbers using the standard algorithm and using estimation to check for reasonableness of the product.
© 2018 Great Minds®. eureka-math.org

EUREKA MATH

4. So far, Carmella has collected 14 boxes of baseball cards. There are 315 cards in each box. Carmella estimates that she has about 3,000 cards, so she buys 6 albums that hold 500 cards each.

    a. Will the albums have enough space for all of her cards? Why or why not?

    b. How many cards does Carmella have?

    c. How many albums will she need for all of her baseball cards?

EUREKA
MATH®

Lesson 8:   Fluently multiply multi-digit whole numbers using the standard
                  algorithm and using estimation to check for reasonableness of the
                  product.
© 2018 Great Minds®. eureka-math.org

175

Name _____   Date _____

Estimate the product first. Solve by using the standard algorithm. Use your estimate to check the reasonableness of the product.

a.   283 × 416

≈ _____ × _____

= _____

$$
\begin{array}{r}
2\,8\,3 \\
\times\ \ 4\,1\,6 \\
\hline
\end{array}
$$

b.   2,803 × 406

≈ _____ × _____

= _____

$$
\begin{array}{r}
2\,,8\,0\,3 \\
\times\ \ \ 4\,0\,6 \\
\hline
\end{array}
$$

Name _____   Date _____

Solve.

1. An office space in New York City measures 48 feet by 56 feet. If it sells for $565 per square foot, what is the total cost of the office space?

2. Gemma and Leah are both jewelry makers. Gemma made 106 beaded necklaces. Leah made 39 more necklaces than Gemma.

    a. Each necklace they make has exactly 104 beads on it. How many beads did both girls use altogether while making their necklaces?

    b. At a recent craft fair, Gemma sold each of her necklaces for $14. Leah sold each of her necklaces for $10 more. Who made more money at the craft fair? How much more?

Lesson 9:   Fluently multiply multi-digit whole numbers using the standard
                  algorithm to solve multi step word problems.

© 2018 Great Minds®. eureka-math.org

179

3. Peng bought 26 treadmills for her new fitness center at $1,334 each. Then, she bought 19 stationary bikes for $749 each. How much did she spend on her new equipment? Write an expression, and then solve.

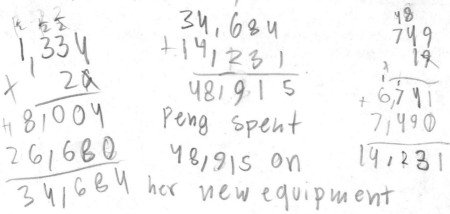

4. A Hudson Valley farmer has 26 employees. He pays each employee $410 per week. After paying his workers for one week, the farmer has $162 left in his bank account. How much money did he have at first?

5. Frances is sewing a border around 2 rectangular tablecloths that each measure 9 feet long by 6 feet wide. If it takes her 3 minutes to sew on 1 inch of border, how many minutes will it take her to complete her sewing project? Write an expression, and then solve.

            Lesson 9:   Fluently multiply multi-digit whole numbers using the standard algorithm to solve multi step word problems.

6) Each grade level at Hooperville Schools has 298 students.

   a. If there are 13 grade levels, how many students attend Hooperville Schools?

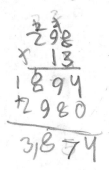

   3854 students
   attend Hooperville
   School

   b. A nearby district, Willington, is much larger. They have 12 times as many students. How many students attend schools in Willington?

   46,488
   attend
   Schools in
   Willington

EUREKA MATH

Lesson 9:   Fluently multiply multi-digit whole numbers using the standard
             algorithm to solve multi step word problems.

© 2018 Great Minds®. eureka-math.org

181

Name _____   Date _____

Solve.

Juwad picked 30 bags of apples on Monday and sold them at his fruit stand for $3.45 each. The following week he picked and sold 26 bags.

   a. How much money did Juwad earn in the first week?

   b. How much money did he earn in the second week?

   c. How much did Juwad earn selling bags of apples these two weeks?

   d. **Extension:** Each bag Juwad picked holds 15 apples. How many apples did he pick in two weeks? Write an expression to represent this problem.

Lesson 9:   Fluently multiply multi-digit whole numbers using the standard
algorithm to solve multi step word problems.

© 2018 Great Minds®. eureka-math.org

183

The fifth-grade craft club is making aprons to sell. Each apron takes 1.25 yards of fabric that costs $3 per yard and 4.5 yards of trim that costs $2 per yard. What does it cost the club to make one apron? If the club wants to make $1.75 profit on each apron, how much should they charge per apron?

_____

_____

_____

_____

Read   Draw   Write

Lesson 10:   Multiply decimal fractions with tenths by multi-digit whole numbers
using place value understanding to record partial products.

185

EUREKA
MATH

© 2018 Great Minds®. eureka-math.org

Name _____ Date _____

1. Estimate the product. Solve using an area model and the standard algorithm. Remember to express your products in standard form.

    a. 22 × 2.4 ≈ _____ × _____ = _____

    2 4 (tenths)
    × 2 2
    ‾‾‾‾‾‾‾

    b. 3.1 × 33 _____ × _____ = _____

    3 1 (tenths)
    × 3 3
    ‾‾‾‾‾‾‾

2. Estimate. Then, use the standard algorithm to solve. Express your products in standard form.

    a. 3.2 × 47 ≈ _____ × _____ = _____

    3 2 (tenths)
    × 4 7
    ‾‾‾‾‾‾‾

    b. 3.2 × 94 ≈ _____ × _____ = _____

    3 2 (tenths)
    × 9 4
    ‾‾‾‾‾‾‾

EUREKA MATH

Lesson 10: Multiply decimal fractions with tenths by multi-digit whole numbers using place value understanding to record partial products.

187

© 2018 Great Minds®. eureka-math.org

c. 6.3 × 44 ≈ _____ × _____ = _____   d. 14.6 × 17 ≈ _____ × _____ = _____

e. 8.2 × 34 ≈ _____ × _____ = _____   f. 160.4 × 17 ≈ _____ × _____ = _____

3. Michelle multiplied 3.4 × 52. She incorrectly wrote 1,768 as her product. Use words, numbers, and/or pictures to explain Michelle's mistake.

4. A wire is bent to form a square with a perimeter of 16.4 cm. How much wire would be needed to form 25 such squares? Express your answer in meters.

Lesson 10:   Multiply decimal fractions with tenths by multi-digit whole numbers using place value understanding to record partial products.

EUREKA MATH

Name _____   Date _____

1. Estimate the product. Solve using an area model and the standard algorithm. Remember to express your products in standard form.

(a.) $33.2 \times 21 \approx$ ___300___ $\times$ ___2___ = ___60___

	300	30	2
1	300	30	2	= 332 tenths
20	6000	600	40	= 6,640 tents

$$\begin{array}{r} 6,640 \\ +\ \ 332 \\ \hline 6,972 = 697.2 \end{array}$$

(b.) $1.7 \times 55 \approx$ ___20___ $\times$ ___60___ = ___1,200___

	10	7
5	50	35	= 85 tenths
50	500	350	= 850 tenths

$$\begin{array}{r} 1 \\ 850 \\ +\ \ 85 \\ \hline 935 = 93.5 \end{array}$$

2. If the product of 485 × 35 is 16,975, what is the product of 485 × 3.5? How do you know?

EUREKA MATH®   Lesson 10:   Multiply decimal fractions with tenths by multi-digit whole numbers using place value understanding to record partial products.   189

© 2018 Great Minds®. eureka-math.org

Mr. Mohr wants to build a rectangular patio using concrete tiles that are 12 square inches. The patio will measure 13.5 feet by 43 feet. What is the area of the patio? How many concrete tiles will he need to complete the patio?

_____

_____

_____

_____

Read   **Draw**   **Write**

Name _____   Date _____

1. Estimate the product. Solve using the standard algorithm. Use the thought bubbles to show your thinking. (Draw an area model on a separate sheet if it helps you.)

a. $1.38 \times 32 \approx$ _____ × _____ = _____ $1.38 \times 32 =$ _____

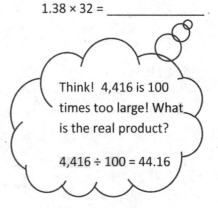

Think!
$1.38 \times 100 = 138$

$$\begin{array}{r} 1.38 \\ \times\ \ 32 \\ \hline \end{array}$$

Think! 4,416 is 100 times too large! What is the real product?

$4,416 \div 100 = 44.16$

b. $3.55 \times 89 \approx$ _____ × _____ = _____ $3.55 \times 89 =$ _____

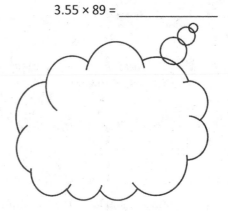

$$\begin{array}{r} 3.55 \\ \times\ \ 89 \\ \hline \end{array}$$

2. Solve using the standard algorithm.

   a. 5.04 × 8   b.   147.83 × 67

   c. 83.41 × 504   d.   0.56 × 432

3. Use the whole number product and place value reasoning to place the decimal point in the second product. Explain how you know.

   a. If 98 × 768 = 75,264   then   98 × 7.68 = _____

   b. If 73 × 1,563 = 114,099   then   73 × 15.63 = _____

   c. If 46 × 1,239 = 56,994   then   46 × 123.9 = _____

Lesson 11:   Multiply decimal fractions by multi-digit whole numbers through
               conversion to a whole number problem and reasoning about the
               placement of the decimal.
© 2018 Great Minds®. eureka-math.org

4. Jenny buys 22 pens that cost $1.15 each and 15 markers that cost $2.05 each. How much did Jenny spend?

5. A living room measures 24 feet by 15 feet. An adjacent square dining room measures 13 feet on each side. If carpet costs $6.98 per square foot, what is the total cost of putting carpet in both rooms?

Name _Fernanda-Arreaza_   Date _____

Use estimation and place value reasoning to find the unknown product. Explain how you know.

1.   If 647 × 63 = 40,761   then   6.47 × 63 = __407.61__

I know this
because I'm
Mulitplying by
hundredths.

$$
\begin{array}{r}
\overset{2\ 4}{647} \\
\times\ \ 63 \\
\hline
+1941 \\
38820 \\
\hline
407.61
\end{array}
$$

2.   Solve using the standard algorithm.

a.   6.13 × 14

$$
\begin{array}{r}
\overset{1}{613} \\
\times\ \ 14 \\
\hline
+2452 \\
6130 \\
\hline
85.82
\end{array}
$$

b.   104.35 × 34

$$
\begin{array}{r}
\overset{1\ 1\ 1}{10435} \\
\times\ \ 34 \\
\hline
+41740 \\
313050 \\
\hline
3547.90
\end{array}
$$

EUREKA
MATH®

Lesson 11:   Multiply decimal fractions by multi-digit whole numbers through
conversion to a whole number problem and reasoning about the
placement of the decimal.
© 2018 Great Minds®. eureka-math.org

197

Thirty-two cyclists make a seven-day trip. Each cyclist requires 8.33 kilograms of food for the entire trip. If each cyclist wants to eat an equal amount of food each day, how many kilograms of food will the group be carrying at the end of Day 5?

_____

_____

_____

_____

Read   **Draw**   **Write**

Lesson 12:   Reason about the product of a whole number and a decimal with
hundredths using place value understanding and estimation.

199

Name _____ Date _____

1. Estimate. Then, solve using the standard algorithm. You may draw an area model if it helps you.

    a. $1.21 \times 14 \approx$ _____ $\times$ _____ = _____

$$\begin{array}{r} 1.21 \\ \times \ \ 14 \\ \hline \end{array}$$

    b. $2.45 \times 305 \approx$ _____ $\times$ _____ = _____

$$\begin{array}{r} 2.45 \\ \times \ 305 \\ \hline \end{array}$$

EUREKA
MATH

Lesson 12: Reason about the product of a whole number and a decimal with
hundredths using place value understanding and estimation.

© 2018 Great Minds®. eureka-math.org

201

2. Estimate. Then, solve using the standard algorithm. Use a separate sheet to draw the area model if it helps you.

a.   1.23 × 12 ≈ _____ × _____ = _____   b.   1.3 × 26 ≈ _____ × _____ = _____

c.   0.23 × 14 ≈ _____ × _____ = _____   d.   0.45 × 26 ≈ _____ × _____ = _____

e.   7.06 × 28 ≈ _____ × _____ = _____   f.   6.32 × 223 ≈ _____ × _____ = _____

g.   7.06 × 208 ≈ _____ × _____ = _____   h.   151.46 × 555 ≈ _____ × _____ = _____

Lesson 12:   Reason about the product of a whole number and a decimal with hundredths using place value understanding and estimation.

© 2018 Great Minds®. eureka-math.org

EUREKA MATH

3. Denise walks on the beach every afternoon. In the month of July, she walked 3.45 miles each day. How far did Denise walk during the month of July?

4. A gallon of gas costs $4.34. Greg puts 12 gallons of gas in his car. He has a 50-dollar bill. Tell how much money Greg will have left, or how much more money he will need. Show all your calculations.

5. Seth drinks a glass of orange juice every day that contains 0.6 grams of Vitamin C. He eats a serving of strawberries for snack after school every day that contains 0.35 grams of Vitamin C. How many grams of Vitamin C does Seth consume in 3 weeks?

Lesson 12:   Reason about the product of a whole number and a decimal with
             hundredths using place value understanding and estimation.

© 2018 Great Minds®. eureka-math.org

203

Name _____ Date _____

Estimate. Then, solve using the standard algorithm.

a. $3.03 \times 402 \approx$ ___3___ $\times$ __400__ = __1200__

$$
\begin{array}{r}
1 \\
303 \\
\times\ 402 \\
\hline
606 \\
+\ 121200 \\
\hline
1218.06
\end{array}
$$

b. $667 \times 1.25 \approx$ __700__ $\times$ __1__ = __700__

$$
\begin{array}{r}
11 \\
667 \\
\times\ 125 \\
\hline
3335 \\
+\ 13340 \\
66700 \\
\hline
833.75
\end{array}
$$

Lesson 12:   Reason about the product of a whole number and a decimal with hundredths using place value understanding and estimation.

205

© 2018 Great Minds®. eureka-math.org

a. Measure your string and express the measurement in meters, centimeters, and millimeters. Record your results in the table in Row A.

	m	cm	mm
A			
B			

b. Measure your partner's string and record the results in the table in Row B.

c. How does the unit of measurement affect the length of the string?

d. Using the numbers on your table, show how to convert from meters to millimeters.

Read   **Draw**   **Write**

Lesson 13:   Use whole number multiplication to express equivalent measurements.

© 2018 Great Minds®. eureka-math.org

207

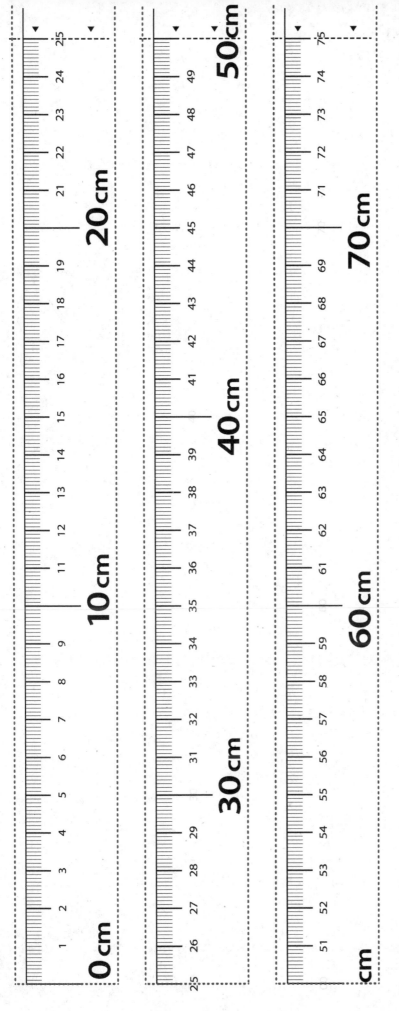

meter strip

Lesson 13: Use whole number multiplication to express equivalent measurements.

© 2018 Great Minds®. eureka-math.org

EUREKA MATH

Name_____   Date _____

1. Solve. The first one is done for you.

a. Convert weeks to days.	b. Convert years to days.
8 weeks = 8 × (1 week) = 8 × (7 days) = 56 days	4 years = _____ × (_____ year) = _____ × (_____ days) = _____ days
c. Convert meters to centimeters. 9.2 m = _____ × (_____ m) = _____ × (_____ cm) = _____ cm	d. Convert yards to feet. 5.7 yards
e. Convert kilograms to grams. 6.08 kg	f. Convert pounds to ounces. 12.5 pounds

2. After solving, write a statement to express each conversion. The first one is done for you.

a. Convert the number of hours in a day to minutes.
 	
    24 hours = 24 × (1 hour)
                = 24 × (60 minutes)
                = 1,440 minutes
 	
One day has 24 hours, which is the same as 1,440 minutes.
	b. A small female gorilla weighs 68 kilograms. How much does she weigh in grams?

c. The height of a man is 1.7 meters. What is his height in centimeters?	d. The capacity of a syringe is 0.08 liters. Convert this to milliliters.
e. A coyote weighs 11.3 pounds. Convert the coyote's weight to ounces.	f. An alligator is 2.3 yards long. What is the length of the alligator in inches?

Lesson 13:   Use whole number multiplication to express equivalent measurements.

© 2018 Great Minds®. eureka-math.org

EUREKA
MATH

Name _____   Date _____

Solve.

a. Convert pounds to ounces. (1 pound = 16 ounces)	b. Convert kilograms to grams.
14 pounds = _____ × (1 pound)	18.2 kilograms = _____ × (_____)
= _____ × (_____ ounces)	= _____ × (_____)
= _____ ounces	= _____ grams

EUREKA MATH   **Lesson 13:**   Use whole number multiplication to express equivalent   **213**
                                        measurements.

© 2018 Great Minds®. eureka-math.org

Draw and label a tape diagram to represent each of the following:

1. Express 1 day as a fraction of 1 week.

2. Express 1 foot as a fraction of 1 yard.

3. Express 1 quart as a fraction of 1 gallon.

Read   **Draw**   **Write**

 EUREKA MATH   **Lesson 14:**   Use fraction and decimal multiplication to express equivalent   215
measurements.

© 2018 Great Minds®. eureka-math.org

4. Express 1 centimeter as a fraction of 1 meter.   (Decimal form.)

5. Express 1 meter as a fraction of 1 kilometer.   (Decimal form.)

Read   **Draw**   **Write**

Lesson 14:   Use fraction and decimal multiplication to express equivalent measurements.

EUREKA MATH

Name _____   Date _____

1. Solve. The first one is done for you.

a. Convert days to weeks.	b. Convert quarts to gallons.
28 days = 28 × (1 day) $= 28 \times \left(\frac{1}{7} \text{ week}\right)$ $= \frac{28}{7}$ week = 4 weeks	20 quarts = _____ × (1 quart) $= \underline{\qquad} \times \left(\frac{1}{4} \text{ gallon}\right)$ = _____ gallons = _____ gallons
c. Convert centimeters to meters.	d. Convert meters to kilometers.
920 cm = _____ × (_____ cm) = _____ × (_____ m) = _____ m	1,578 m = _____ × (_____ m) = _____ × (0.001 km) = _____ km
e. Convert grams to kilograms.	f. Convert milliliters to liters.
6,080 g =	509 mL =

2. After solving, write a statement to express each conversion. The first one is done for you.

a. The screen measures 24 inches. Convert 24 inches to feet. 24 inches = 24 × (1 inch) $= 24 \times \left(\frac{1}{12} \text{ feet}\right)$ $= \frac{24}{12} \text{ feet}$ = 2 feet The screen measures 24 inches or 2 feet.	b. A jug of syrup holds 12 cups. Convert 12 cups to pints.
c. The length of the diving board is 378 centimeters. What is its length in meters?	d. The capacity of a container is 1,478 milliliters. Convert this to liters.
e. A truck weighs 3,900,000 grams. Convert the truck's weight to kilograms.	f. The distance was 264,040 meters. Convert the distance to kilometers.

Lesson 14: Use fraction and decimal multiplication to express equivalent measurements.

© 2018 Great Minds®. eureka-math.org

EUREKA MATH

Name _____   Date _____

1. Convert days to weeks by completing the number sentences.

   35 days = _____ × (_____ day)

         = _____ × (_____ week)

         .=

         =

2. Convert grams to kilograms by completing the number sentences.

   4,567 grams = _____ × _____

             = _____ × _____

             .=

             =

EUREKA
MATH®

Lesson 14:   Use fraction and decimal multiplication to express equivalent
             measurements.

© 2018 Great Minds®. eureka-math.org

219

Name _____   Date _____

Solve.

1. Liza's cat had six kittens! When Liza and her brother weighed all the kittens together, they weighed 4 pounds 2 ounces. Since all the kittens are about the same size, about how many ounces does each kitten weigh?

2. A container of oregano is 17 pounds heavier than a container of peppercorns. Their total weight is 253 pounds. The peppercorns will be sold in one-ounce bags. How many bags of peppercorns can be made?

Lesson 15:   Solve two-step word problems involving measurement conversions.

221

© 2018 Great Minds®. eureka-math.org

3. Each costume needs 46 centimeters of red ribbon and 3 times as much yellow ribbon. What is the total length of ribbon needed for 64 costumes? Express your answer in meters.

4. When making a batch of orange juice for her basketball team, Jackie used 5 times as much water as concentrate. There were 32 more cups of water than concentrate.

    a. How much juice did she make in all?

    b. She poured the juice into quart containers. How many containers could she fill?

Lesson 15:   Solve two-step word problems involving measurement conversions.

© 2018 Great Minds®. eureka-math.org

Name _____   Date _____

Solve.

To practice for an Ironman competition, John swam 0.86 kilometer each day for 3 weeks. How many meters did he swim in those 3 weeks?

The area of a rectangular vegetable garden is 200 ft². The width is 10 ft. What is the length of the vegetable garden?

_____

_____

_____

_____

Read   **Draw**   **Write**

Lesson 16:   Use *divide by 10* patterns for multi-digit whole number division.

225

© 2018 Great Minds®. eureka-math.org

Name _____ Date _____

1. Divide. Draw place value disks to show your thinking for (a) and (c). You may draw disks on your personal white board to solve the others if necessary.

a. $500 \div 10 = 50$	b. $360 \div 10$
c. $12,000 \div 100 = 120$	d. $450,000 \div 100$
e. $700,000 \div 1,000 = 700$	f. $530,000 \div 100$

Lesson 16: Use *divide by 10* patterns for multi-digit whole number division.

227

© 2018 Great Minds®. eureka-math.org

2. Divide. The first one is done for you.

a. 12,000 ÷ 30	b. 12,000 ÷ 300	c. 12,000 ÷ 3,000
= 12,000 ÷ 10 ÷ 3 = 1,200 ÷ 3 = 400	= 12,000 ÷ 100 = 120 ÷ 3 = 40	
d. 560,000 ÷ 70	e. 560,000 ÷ 700	f. 560,000 ÷ 7,000
= 560,000 ÷ 10 = 56,000 ÷ 7 = 8,000		
g. 28,000 ÷ 40	h. 450,000 ÷ 500	i. 810,000 ÷ 9,000
		= 810,000 ÷ 1000 = 810 ÷ 9 = 90

Lesson 16: Use *divide by 10* patterns for multi-digit whole number division.

3. The floor of a rectangular banquet hall has an area of 3,600 m². The length is 90 m.

   a. What is the width of the banquet hall?

   b. A square banquet hall has the same area. What is the length of the room?

   c. A third rectangular banquet hall has a perimeter of 3,600 m. What is the width if the length is 5 times the width?

Lesson 16: Use *divide by 10* patterns for multi-digit whole number division.

229

© 2018 Great Minds®. eureka-math.org

4. Two fifth graders solved 400,000 divided by 800. Carter said the answer is 500, while Kim said the answer is 5,000.

    a. Who has the correct answer? Explain your thinking.

    b. What if the problem is 4,000,000 divided by 8,000? What is the quotient?

Lesson 16:   Use *divide by 10* patterns for multi-digit whole number division.

Name _____   Date _____

Divide. Show your thinking.

a. 17,000 ÷ 100	b. 59,000 ÷ 1,000
= 17000 ÷ 100 = 170	= 59,000 ÷ 1000 = 59
c. 12,000 ÷ 40	d. 480,000 ÷ 600
= 12,000 ÷ 10 = 1200 ÷ 4 = 300	= 480,000 ÷ 100 = 4 800 ÷ 6 = 800

852 pounds of grapes were packed equally into 3 boxes for shipping. How many pounds of grapes were there in 2 boxes?

_____

_____

_____

_____

Read   **Draw**   **Write**

Name __Fernanda-Arreaza__   Date _____

1. Estimate the quotient for the following problems. Round the divisor first.

a. $609 \div 21$ $\approx 600 \div 20$ $= 30$	b. $913 \div 29$ $\approx \underline{900} \div \underline{30}$ $= \underline{30}$	c. $826 \div 37$ $\approx \underline{800} \div \underline{40}$ $= \underline{20}$
d. $141 \div 73$ $\approx \underline{140} \div \underline{70}$ $= \underline{2}$	e. $241 \div 58$ $\approx \underline{240} \div \underline{60}$ $= \underline{4}$	f. $482 \div 62$ $\approx \underline{500} \div \underline{60}$ $= \underline{12}$
g. $656 \div 81$ $\approx \underline{700} \div \underline{80}$ $= \underline{}$	h. $799 \div 99$ $\approx \underline{} \div \underline{}$ $= \underline{}$	i. $635 \div 95$ $\approx \underline{} \div \underline{}$ $= \underline{}$
j. $311 \div 76$ $\approx \underline{} \div \underline{}$ $= \underline{}$	k. $648 \div 83$ $\approx \underline{} \div \underline{}$ $= \underline{}$	l. $143 \div 35$ $\approx \underline{} \div \underline{}$ $= \underline{}$
m. $525 \div 25$ $\approx \underline{} \div \underline{}$ $= \underline{}$	n. $552 \div 85$ $\approx \underline{} \div \underline{}$ $= \underline{}$	o. $667 \div 11$ $\approx \underline{} \div \underline{}$ $= \underline{}$

Lesson 17: Use basic facts to approximate quotients with two-digit divisors.

235

EUREKA MATH®

2. A video game store has a budget of $825, and would like to purchase new video games. If each video game costs $41, estimate the total number of video games the store can purchase with its budget. Explain your thinking.

3. Jackson estimated 637 ÷ 78 as 640 ÷ 80. He reasoned that 64 tens divided by 8 tens should be 8 tens. Is Jackson's reasoning correct? If so, explain why. If not, explain a correct solution.

EUREKA
MATH®

Name _____   Date _____

Estimate the quotient for the following problems.

a. $608 \div 23$ $\approx$ $\underline{600}$ $\div$ $\underline{20}$ $=$ $\underline{30}$	b. $913 \div 31$ $\approx$ $\underline{900}$ $\div$ $\underline{30}$ $=$ $\underline{30}$
c. $151 \div 39$ $\approx$ $\underline{200}$ $\div$ $\underline{40}$ $=$ $\underline{5}$	d. $481 \div 68$ $\approx$ _____ $\div$ _____ $=$ _____

$4 \overline{)20}$

$4 \times 5 = 20$

$\begin{array}{r} 40 \\ \times\ 5 \\ \hline 200 \end{array}$

EUREKA MATH®

Sandra bought 38 DVD movies for $874. Give an estimate of the cost of each DVD movie.

_____

_____

_____

_____

Read   **Draw**   **Write**

Lesson 18:   Use basic facts to approximate quotients with two-digit divisors.

239

Name _____   Date _____

1. Estimate the quotients for the following problems. The first one is done for you.

a. 5,738 ÷ 21 ≈ 6,000 ÷ 20 = 300	b. 2,659 ÷ 28 ≈ _____ ÷ _____ = _____	c. 9,155 ÷ 34 ≈ _____ ÷ _____ = _____
d. 1,463 ÷ 53 ≈ _____ ÷ _____ = _____	e. 2,525 ÷ 64 ≈ _____ ÷ _____ = _____	f. 2,271 ÷ 72 ≈ _____ ÷ _____ = _____
g. 4,901 ÷ 75 ≈ _____ ÷ _____ = _____	h. 8,515 ÷ 81 ≈ _____ ÷ _____ = _____	i. 8,515 ÷ 89 ≈ _____ ÷ _____ = _____
j. 3,925 ÷ 68 ≈ _____ ÷ _____ = _____	k. 5,124 ÷ 81 ≈ _____ ÷ _____ = _____	l. 4,945 ÷ 93 ≈ _____ ÷ _____ = _____
m. 5,397 ÷ 94 ≈ _____ ÷ _____ = _____	n. 6,918 ÷ 86 ≈ _____ ÷ _____ = _____	o. 2,806 ÷ 15 ≈ _____ ÷ _____ = _____

2. A swimming pool requires 672 ft² of floor space. The length of the swimming pool is 32 ft. Estimate the width of the swimming pool.

3. Janice bought 28 apps for her phone that, altogether, used 1,348 MB of space.

   a. If each app used the same amount of space, about how many MB of memory did each app use? Show how you estimated.

   b. If half of the apps were free and the other half were $1.99 each, about how much did she spend?

4. A quart of paint covers about 85 square feet. About how many quarts would you need to cover a fence with an area of 3,817 square feet?

5. Peggy has saved $9,215. If she is paid $45 an hour, about how many hours did she work?

Lesson 18:   Use basic facts to approximate quotients with two-digit divisors.

Name __Fernanda_____   Date _____

Estimate the quotients for the following problems.

a. $6{,}523 \div 21$	b. $8{,}491 \div 37$
$\approx$ _6000_ $\div$ _20_ $=$ _300_	$\approx$ _8000_ $\div$ _40_ $=$ _200_
c. $3{,}704 \div 53$	d. $4{,}819 \div 68$
$\approx$ _4000_ $\div$ _50_ $=$ _800_	$\approx$ _4,900_ $\div$ _70_ $=$ _700_

At the Highland Falls pumpkin-growing contest, the prize winning pumpkin contains 360 seeds. The proud farmer plans to sell his seeds in packs of 12. How many packs can he make using all the seeds?

_____

_____

_____

_____

Read　　　　**Draw**　　　　**Write**

Lesson 19:　Divide two- and three-digit dividends by multiples of 10 with single-digit quotients, and make connections to a written method.

245

© 2018 Great Minds®. eureka-math.org

Name _____   Date _____

1. Divide, and then check. The first problem is done for you.

   a. 41 ÷ 30

```
              1   R 11
      3  0 | 4   1
         -   3   0
             1   1
```

   Check:

   30 × 1 = 30
   30 + 11 = 41

   b. 80 ÷ 30

   c. 71 ÷ 50

   d. 270 ÷ 30

   e. 643 ÷ 80

   f. 215 ÷ 90

EUREKA MATH

Lesson 19: Divide two- and three-digit dividends by multiples of 10 with single-digit quotients, and make connections to a written method.

247

© 2018 Great Minds®. eureka-math.org

2. Terry says the solution to 299 ÷ 40 is 6 with a remainder of 59. His work is shown below. Explain Terry's error in thinking, and then find the correct quotient using the space on the right.

```
            6
  4 0 | 2 9 9                   4 0 | 2 9 9
    -  2 4 0
          5 9
```

3. A number divided by 80 has a quotient of 7 with 4 as a remainder. Find the number.

4. While swimming a 2 km race, Adam changes from breaststroke to butterfly every 200 m. How many times does he switch strokes during the first half of the race?

Lesson 19: Divide two- and three-digit dividends by multiples of 10 with single-digit quotients, and make connections to a written method.

© 2018 Great Minds®. eureka-math.org

EUREKA
MATH

Name _____   Date _____

Divide, and then check using multiplication.

    a. $73 \div 20$

    b. $291 \div 30$

Lesson 19:   Divide two- and three-digit dividends by multiples of 10 with
single-digit quotients, and make connections to a written method.

249

© 2018 Great Minds®. eureka-math.org

Billy has 2.4 m of ribbon for crafts. He wants to share it evenly with 12 friends. How many centimeters of ribbon would 7 friends get?

_____

_____

_____

_____

Read   **Draw**   **Write**

Lesson 20:   Divide two- and three-digit dividends by two-digit divisors with single digit quotients, and make connections to a written method.

251

© 2018 Great Minds®. eureka-math.org

Name _____   Date _____

1. Divide. Then, check with multiplication. The first one is done for you.

   a.   65 ÷ 17

$$\begin{array}{r} 3 \text{ R } 14 \\ 17 \overline{\smash{\big)}\, 65} \\ -\underline{51} \\ 14 \end{array}$$

   Check:

   17 × 3 = 51

   51 + 14 = 65

   b.   49 ÷ 21

   c.   78 ÷ 39

   d.   84 ÷ 32

   e.   77 ÷ 25

   f.   68 ÷ 17

Lesson 20: Divide two- and three-digit dividends by two-digit divisors with single digit quotients, and make connections to a written method.

253

2. When dividing 82 by 43, Linda estimated the quotient to be 2. Examine Linda's work, and explain what she needs to do next. On the right, show how you would solve the problem.

Linda's Estimation:	Linda's Work:	Your Work:

$$
\begin{array}{r}
2 \\
40\,\overline{)\,8\ 0}
\end{array}
$$

$$
\begin{array}{r}
2 \\
43\,\overline{)\,8\ 2} \\
-\ 8\ 6 \\
\hline
?\ ?
\end{array}
$$

$$
43\,\overline{)\,8\ 2}
$$

3. A number divided by 43 has a quotient of 3 with 28 as a remainder. Find the number. Show your work.

Lesson 20:   Divide two- and three-digit dividends by two-digit divisors with single digit quotients, and make connections to a written method.

© 2018 Great Minds®. eureka-math.org

EUREKA MATH

4. Write another division problem that has a quotient of 3 and a remainder of 28.

5. Mrs. Silverstein sold 91 cupcakes at a food fair. The cupcakes were sold in boxes of "a baker's dozen," which is 13. She sold all the cupcakes at $15 per box. How much money did she receive?

Lesson 20:   Divide two- and three-digit dividends by two-digit divisors with single digit quotients, and make connections to a written method.

© 2018 Great Minds®. eureka-math.org

255

Name _____ Date _____

Divide. Then, check with multiplication.

    a. $78 \div 21$

    b. $89 \div 37$

Lesson 20:   Divide two- and three-digit dividends by two-digit divisors with single digit quotients, and make connections to a written method.

257

© 2018 Great Minds®. eureka-math.org

105 students were divided equally into 15 teams.

    a. How many players were on each team?

    b. If each team had 3 girls, how many boys were there altogether?

Read   **Draw**   **Write**

Lesson 21:   Divide two- and three-digit dividends by two-digit divisors with single digit quotients, and make connections to a written method.

259

© 2018 Great Minds®. eureka-math.org

Name _____   Date _____

1. Divide. Then, check using multiplication. The first one is done for you.

   a. 258 ÷ 47

   ```
              5 R 23
      4 7 | 2 5 8
        -   2 3 5
            ─────
              2 3
   ```

   Check:

   47 × 5 = 235

   235 + 23 = 258

   b. 148 ÷ 67

   c. 591 ÷ 73

   d. 759 ÷ 94

Lesson 21:   Divide two- and three-digit dividends by two-digit divisors with single
digit quotients, and make connections to a written method.

© 2018 Great Minds®. eureka-math.org

261

e. $653 \div 74$

f. $257 \div 36$

2. Generate and solve at least one more division problem with the same quotient and remainder as the one below. Explain your thought process.

```
            8
  58 | 4 7 5
   -   4 6 4
           1 1
```

Divide two- and three-digit dividends by two-digit divisors with single
              digit quotients, and make connections to a written method.

EUREKA
MATH®

3. Assume that Mrs. Giang's car travels 14 miles on each gallon of gas. If she travels to visit her niece who lives 133 miles away, how many gallons of gas will Mrs. Giang need to make the round trip?

4. Louis brings 79 pencils to school. After he gives each of his 15 classmates an equal number of pencils, he will give any leftover pencils to his teacher.

   a. How many pencils will Louis's teacher receive?

   b. If Louis decides instead to take an equal share of the pencils along with his classmates, will his teacher receive more pencils or fewer pencils? Show your thinking.

 EUREKA MATH   **Lesson 21:**   Divide two- and three-digit dividends by two-digit divisors with single digit quotients, and make connections to a written method.   **263**

© 2018 Great Minds®. eureka-math.org

Name _____   Date _____

Divide. Then, check using multiplication.

a. $326 \div 53$

b. $192 \div 38$

Lesson 21:   Divide two- and three-digit dividends by two-digit divisors with single
digit quotients, and make connections to a written method.

© 2018 Great Minds®. eureka-math.org

265

Zenin's baby sister weighed 132 ounces at birth. How much did his sister weigh in pounds and ounces?

Read   **Draw**   **Write**

Lesson 22:   Divide three- and four-digit dividends by two-digit divisors resulting in two- and three-digit quotients, reasoning about the decomposition of successive remainders in each place value.

© 2018 Great Minds®. eureka-math.org

267

Name _____ Date _____

1. Divide. Then, check using multiplication. The first one is done for you.

    a. 580 ÷ 17

    $$
    \begin{array}{r}
    3\ 4\ R\,2 \\
    17\,\overline{)\,5\ 8\ 0} \\
    -\ 5\ 1 \\
    \hline
    7\ 0 \\
    -\ 6\ 8 \\
    \hline
    2
    \end{array}
    $$

    Check:

    34 × 17 = 578

    578 + 2 = 580

    b. 730 ÷ 32

    c. 940 ÷ 28

    d. 553 ÷ 23

Lesson 22:   Divide three- and four-digit dividends by two-digit divisors resulting in two- and three-digit quotients, reasoning about the decomposition of successive remainders in each place value.

© 2018 Great Minds®. eureka-math.org

269

e. 704 ÷ 46

f. 614 ÷ 15

2. Halle solved 664 ÷ 48 below. She got a quotient of 13 with a remainder of 40. How could she use her work below to solve 659 ÷ 48 without redoing the work? Explain your thinking.

```
        1 3
   48 | 6 6 4
     -  4 8
        1 8 4
     -  1 4 4
          4 0
```

Lesson 22:   Divide three- and four-digit dividends by two-digit divisors resulting in two- and three-digit quotients, reasoning about the decomposition of successive remainders in each place value.

© 2018 Great Minds®. eureka-math.org

3. 27 students are learning to make balloon animals. There are 172 balloons to be shared equally among the students.

    a. How many balloons are left over after sharing them equally?

    b. If each student needs 7 balloons, how many more balloons are needed? Explain how you know.

Lesson 22:   Divide three- and four-digit dividends by two-digit divisors resulting in two- and three-digit quotients, reasoning about the decomposition of successive remainders in each place value.

© 2018 Great Minds®. eureka-math.org

271

Name _____   Date _____

Divide. Then, check using multiplication.

    a. $413 \div 19$

    b. $708 \div 67$

Lesson 22:   Divide three- and four-digit dividends by two-digit divisors resulting in
two- and three-digit quotients, reasoning about the decomposition of
successive remainders in each place value.

© 2018 Great Minds®. eureka-math.org

273

The rectangular room measures 224 square feet. One side of the room is 14 feet long. What is the perimeter of the room?

_____

_____

_____

_____

Read   Draw   Write

Lesson 23: Divide three- and four-digit dividends by two-digit divisors resulting in two- and three-digit quotients, reasoning about the decomposition of successive remainders in each place value.

© 2018 Great Minds®. eureka-math.org

275

Name _____   Date _____

1. Divide. Then, check using multiplication.

   a. $4,859 \div 23$ b. $4,368 \div 52$

   c. $7,242 \div 34$ d. $3,164 \div 45$

   e. $9,152 \div 29$ f. $4,424 \div 63$

EUREKA MATH®

Lesson 23:   Divide three- and four-digit dividends by two-digit divisors resulting in two- and three-digit quotients, reasoning about the decomposition of successive remainders in each place value.

© 2018 Great Minds®. eureka-math.org

277

2. Mr. Riley baked 1,692 chocolate cookies. He sold them in boxes of 36 cookies each. How much money did he collect if he sold them all at $8 per box?

3. 1,092 flowers are arranged into 26 vases, with the same number of flowers in each vase. How many flowers would be needed to fill 130 such vases?

4. The elephant's water tank holds 2,560 gallons of water. After two weeks, the zookeeper measures and finds that the tank has 1,944 gallons of water left. If the elephant drinks the same amount of water each day, how many days will a full tank of water last?

Lesson 23:   Divide three- and four-digit dividends by two-digit divisors resulting in
                         two- and three-digit quotients, reasoning about the decomposition of
                         successive remainders in each place value.
            © 2018 Great Minds®. eureka-math.org

Name _____   Date _____

Divide. Then, check using multiplication.

a.   8,283 ÷ 19

b.   1,056 ÷ 37

Lesson 23:   Divide three- and four-digit dividends by two-digit divisors resulting in
two- and three-digit quotients, reasoning about the decomposition of
successive remainders in each place value.

© 2018 Great Minds®. eureka-math.org

279

A long-time runner compiled her training distances in the following chart. Fill in the missing values.

Runner's Log

Total Number of Miles Run	Number of Days	Miles Run Each Day
420		12
14.5	5	
38.0	10	
	17	16.5

Read   **Draw**   **Write**

Lesson 24:   Divide decimal dividends by multiples of 10, reasoning about the placement of the decimal point and making connections to a written method.

© 2018 Great Minds®. eureka-math.org

281

Name _____   Date _____

1. Divide. Show the division in the right-hand column in two steps. The first two have been done for you.

a. $1.2 \div 6 = 0.2$

b. $1.2 \div 60 = (1.2 \div 6) \div 10 = 0.2 \div 10 = 0.02$

c. $2.4 \div 4 =$ _____

d. $2.4 \div 40 =$ _____

e. $14.7 \div 7 =$ _____

f. $14.7 \div 70 =$ _____

g. $0.34 \div 2 =$ _____

h. $3.4 \div 20 =$ _____

i. $0.45 \div 9 =$ _____

j. $0.45 \div 90 =$ _____

k. $3.45 \div 3 =$ _____

l. $34.5 \div 300 =$ _____

2. Use place value reasoning and the first quotient to compute the second quotient. Explain your thinking.

   a. $46.5 \div 5 = 9.3$

   $46.5 \div 50 =$ _____

   b. $0.51 \div 3 = 0.17$

   $0.51 \div 30 =$ _____

   c. $29.4 \div 70 = 0.42$

   $29.4 \div 7 =$ _____

   d. $13.6 \div 40 = 0.34$

   $13.6 \div 4 =$ _____

Lesson 24: Divide decimal dividends by multiples of 10, reasoning about the placement of the decimal point and making connections to a written method.

EUREKA MATH

3. Twenty polar bears live at the zoo. In four weeks, they eat 9,732.8 pounds of food altogether. Assuming each bear is fed the same amount of food, how much food is used to feed one bear for a week? Round your answer to the nearest pound.

4. The total weight of 30 bags of flour and 4 bags of sugar is 42.6 kg. If each bag of sugar weighs 0.75 kg, what is the weight of each bag of flour?

Lesson 24:   Divide decimal dividends by multiples of 10, reasoning about the placement of the decimal point and making connections to a written method.

© 2018 Great Minds®. eureka-math.org

285

Name _____   Date _____

1. Divide.

   a. 27.3 ÷ 3

   b. 2.73 ÷ 30

   c. 273 ÷ 300

2. If 7.29 ÷ 9 = 0.81, then the quotient of 7.29 ÷ 90 is _____. Use place value reasoning to explain the placement of the decimal point.

Lesson 24:   Divide decimal dividends by multiples of 10, reasoning about the placement of the decimal point and making connections to a written method.

© 2018 Great Minds®. eureka-math.org

287

Ms. Heinz spent 12 dollars on 30 bus tokens for the field trip. What was the cost of 12 tokens?

_____

_____

_____

_____

Read   Draw   Write

Lesson 25:   Use basic facts to approximate decimal quotients with two-digit   289
                 divisors, reasoning about the placement of the decimal point.

© 2018 Great Minds®. eureka-math.org

Name _____   Date _____

1. Estimate the quotients.

a. $3.24 \div 82 \approx$

b. $361.2 \div 61 \approx$

c. $7.15 \div 31 \approx$

d. $85.2 \div 31 \approx$

e. $27.97 \div 28 \approx$

2. Estimate the quotient in (a). Use your estimated quotient to estimate (b) and (c).

a. $7.16 \div 36 \approx$

b. $716 \div 36 \approx$

c. $71.6 \div 36 \approx$

EUREKA MATH

Lesson 25: Use basic facts to approximate decimal quotients with two-digit divisors, reasoning about the placement of the decimal point.

© 2018 Great Minds®. eureka-math.org

291

3. Edward bikes the same route to and from school each day. After 28 school days, he bikes a total distance of 389.2 miles.

    a. Estimate how many miles he bikes in one day.

    b. If Edward continues his routine of biking to school, about how many days altogether will it take him to reach a total distance of 500 miles?

4. Xavier goes to the store with $40. He spends $38.60 on 13 bags of popcorn.

    a. About how much does one bag of popcorn cost?

    b. Does he have enough money for another bag? Use your estimate to explain your answer.

Lesson 25:   Use basic facts to approximate decimal quotients with two-digit divisors, reasoning about the placement of the decimal point.

© 2018 Great Minds®. eureka-math.org

EUREKA MATH

Name _____   Date _____

Estimate the quotients.

a. $1.64 \div 22 \approx$

b. $123.8 \div 62 \approx$

c. $6.15 \div 31 \approx$

Lesson 25:   Use basic facts to approximate decimal quotients with two-digit
divisors, reasoning about the placement of the decimal point.

© 2018 Great Minds®. eureka-math.org

293

Find the whole number quotient and remainder of the following two expressions:

$$201 \div 12 \qquad\qquad 729 \div 45$$

Use >, <, or = to complete this sentence:

$$201 \div 12 \ \_\_\_\_\ 729 \div 45$$

Justify your answer using decimal quotients.

Read　　　　**Draw**　　　　**Write**

Lesson 26:　Divide decimal dividends by two-digit divisors, estimating quotients,
reasoning about the placement of the decimal point, and making
connections to a written method.

© 2018 Great Minds®. eureka-math.org

295

Name _____   Date _____

1. 156 ÷ 24 and 102 ÷ 15 both have a quotient of 6 and a remainder of 12.

    a. Are the division expressions equivalent to each other? Use your knowledge of decimal division to justify your answer.

    b. Construct your own division problem with a two-digit divisor that has a quotient of 6 and a remainder of 12 but is not equivalent to the problems in 1(a).

2. Divide. Then, check your work with multiplication.

    a. 36.14 ÷ 13   b. 62.79 ÷ 23

    c. 12.21 ÷ 11   d. 6.89 ÷ 13

e. 249.6 ÷ 52   f. 24.96 ÷ 52

g. 300.9 ÷ 59   h. 30.09 ÷ 59

3. The weight of 72 identical marbles is 183.6 grams. What is the weight of each marble? Explain how you
    know the decimal point of your quotient is placed reasonably.

Lesson 26:   Divide decimal dividends by two-digit divisors, estimating quotients,
                            reasoning about the placement of the decimal point, and making
                            connections to a written method.
                            © 2018 Great Minds®. eureka-math.org

4. Cameron wants to measure the length of his classroom using his foot as a length unit. His teacher tells him the length of the classroom is 23 meters. Cameron steps across the classroom heel to toe and finds that it takes him 92 steps. How long is Cameron's foot in meters?

5. A blue rope is three times as long as a red rope. A green rope is 5 times as long as the blue rope. If the total length of the three ropes is 508.25 meters, what is the length of the blue rope?

EUREKA
MATH®

Lesson 26:   Divide decimal dividends by two-digit divisors, estimating quotients,
reasoning about the placement of the decimal point, and making
connections to a written method.
© 2018 Great Minds®. eureka-math.org

299

Name _____   Date _____

1. Estimate. Then, divide using the standard algorithm and check.

   a. $45.15 \div 21$ b. $14.95 \div 65$

2. We learned today that division expressions that have the same quotient and remainders are not necessarily equal to each other. Explain how this is possible.

Lesson 26:   Divide decimal dividends by two-digit divisors, estimating quotients, reasoning about the placement of the decimal point, and making connections to a written method.

© 2018 Great Minds®. eureka-math.org

301

Michael has 567 pennies, Jorge has 464 pennies, and Jaime has 661 pennies. If the pennies are shared equally by the 3 boys and 33 of their classmates, how much money will each classmate receive? Express your final answer in dollars.

Read　　　　**Draw**　　　　**Write**

Lesson 27:　　Divide decimal dividends by two-digit divisors, estimating quotients, reasoning about the placement of the decimal point, and making connections to a written method.

© 2018 Great Minds®. eureka-math.org

303

Name _____ Date _____

1. Divide. Check your work with multiplication.

    a. $5.6 \div 16$

    b. $21 \div 14$

    c. $24 \div 48$

    d. $36 \div 24$

    e. $81 \div 54$

    f. $15.6 \div 15$

    g. $5.4 \div 15$

    h. $16.12 \div 52$

    i. $2.8 \div 16$

2. 30.48 kg of beef was placed into 24 packages of equal weight. What is the weight of one package of beef?

3. What is the length of a rectangle whose width is 17 inches and whose area is 582.25 in²?

Lesson 27: Divide decimal dividends by two-digit divisors, estimating quotients, reasoning about the placement of the decimal point, and making connections to a written method.

© 2018 Great Minds®. eureka-math.org

4. A soccer coach spent $162 dollars on 24 pairs of socks for his players. How much did five pairs of socks cost?

5. A craft club makes 95 identical paperweights to sell. They collect $230.85 from selling all the paperweights. If the profit the club collects on each paperweight is two times as much as the cost to make each one, what does it cost the club to make each paperweight?

Lesson 27: Divide decimal dividends by two-digit divisors, estimating quotients, reasoning about the placement of the decimal point, and making connections to a written method.

© 2018 Great Minds®. eureka-math.org

307

Name _____ Date _____

Divide.

a. 28 ÷ 32

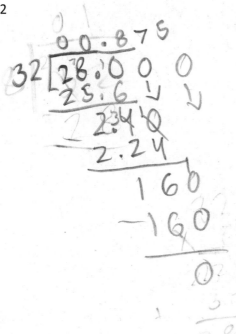

b. 68.25 ÷ 65

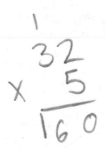

EUREKA MATH®

Lesson 27: Divide decimal dividends by two-digit divisors, estimating quotients, reasoning about the placement of the decimal point, and making connections to a written method.

© 2018 Great Minds®. eureka-math.org

309

Name _____   Date _____

1. Ava is saving for a new computer that costs $1,218. She has already saved half of the money. Ava earns $14.00 per hour. How many hours must Ava work in order to save the rest of the money?

2. Michael has a collection of 1,404 sports cards. He hopes to sell the collection in packs of 36 cards and make $633.75 when all the packs are sold. If each pack is priced the same, how much should Michael charge per pack?

Lesson 28: Solve division word problems involving multi-digit division with group size unknown and the number of groups unknown.

311

© 2018 Great Minds®. eureka-math.org

3. Jim Nasium is building a tree house for his two daughters. He cuts 12 pieces of wood from a board that is 128 inches long. He cuts 5 pieces that measure 15.75 inches each and 7 pieces evenly cut from what is left. Jim calculates that, due to the width of his cutting blade, he will lose a total of 2 inches of wood after making all of the cuts. What is the length of each of the seven pieces?

4. A load of bricks is twice as heavy as a load of sticks. The total weight of 4 loads of bricks and 4 loads of sticks is 771 kilograms. What is the total weight of 1 load of bricks and 3 loads of sticks?

Lesson 28: Solve division word problems involving multi-digit division with group size unknown and the number of groups unknown.

© 2018 Great Minds®. eureka-math.org

EUREKA MATH

Name _____   Date _____

Solve this problem, and show all of your work.

Kenny is ordering uniforms for both the girls' and boys' tennis clubs. He is ordering shirts for 43 players and two coaches at a total cost of $658.35. Additionally, he is ordering visors for each player at a total cost of $368.51. How much will each player pay for the shirt and visor?

Lesson 28:   Solve division word problems involving multi-digit division with group size unknown and the number of groups unknown.   313

© 2018 Great Minds®. eureka-math.org

A one-year (52-week) subscription to a weekly magazine is $39.95. Greg calculates that he would save $219.53 if he subscribed to the magazine instead of purchasing it each week at the store. What is the price of the individual magazine at the store?

_____

_____

_____

_____

Read   **Draw**   **Write**

Lesson 29:   Solve division word problems involving multi-digit division with group
              size unknown and the number of groups unknown.

© 2018 Great Minds®. eureka-math.org

315

Name _____   Date _____

Solve.

1. Lamar has 1,354.5 kilograms of potatoes to deliver equally to 18 stores. 12 of the stores are in the Bronx. How many kilograms of potatoes will be delivered to stores in the Bronx?

2. Valerie uses 12 fluid oz of detergent each week for her laundry. If there are 75 fluid oz of detergent in the bottle, in how many weeks will she need to buy a new bottle of detergent? Explain how you know.

     Lesson 29:   Solve division word problems involving multi-digit division with group   317
size unknown and the number of groups unknown.

© 2018 Great Minds®. eureka-math.org

3. The area of a rectangle is 56.96 m². If the length is 16 m, what is its perimeter?

4. A city block is 3 times as long as it is wide. If the distance around the block is 0.48 kilometers, what is the area of the block in square meters?

Solve division word problems involving multi-digit division with group size unknown and the number of groups unknown.

EUREKA MATH

Name _____ Date _____

Solve.

Hayley borrowed $1,854 from her parents. She agreed to repay them in equal installments throughout the next 18 months. How much will Hayley still owe her parents after a year?

Lesson 29: Solve division word problems involving multi-digit division with group size unknown and the number of groups unknown.

© 2018 Great Minds®. eureka-math.org

319

Credits

Great Minds® has made every effort to obtain permission for the reprinting of all copyrighted material. If any owner of copyrighted material is not acknowledged herein, please contact Great Minds for proper acknowledgment in all future editions and reprints of this module.